Pierre Duhem

Essays in the History and
Philosophy of Science

Translated and Edited, with Introduction, by

ROGER ARIEW and PETER BARKER

ESSAYS IN THE HISTORY AND PHILOSOPHY OF SCIENCE

PIERRE DUHEM

ESSAYS IN THE HISTORY AND PHILOSOPHY OF SCIENCE

Translated and Edited, with Introduction, by
ROGER ARIEW
and
PETER BARKER

Hackett Publishing Company
Indianapolis & Cambridge

Pierre Duhem: 1861–1916

Copyright © 1996 by Hackett Publishing Company, Inc.

All rights reserved
Printed in the United States of America

00 99 98 97 96   1 2 3 4 5

For further information, please address

Hackett Publishing Company, Inc.
P.O. Box 44937
Indianapolis, Indiana 46244-0937

Text design by Dan Kirklin

Library of Congress Cataloging-in-Publication Data

Duhem, Pierre Maurice Marie, 1861–1916.
    Essays in the history and philosophy of science/translated and
edited, with introduction, by Roger Ariew and Peter Barker.
        p.   cm.
    Includes bibliographical references and index.
    ISBN 0-87220-308-5 (pbk.) ISBN 0-87220-309-3 (cloth)
    1. Physics—Philosophy. 2. Science—Philosophy. I. Ariew,
Roger.   II. Barker, Peter, 1949–   .   III. Title.
QC6.2.D84 1996
530'.01—dc20   95-52525
                                                    CIP

Contents

Errata

p. v should read:

6. From *To Save the Phenomena*: Essay on the Concept of
    Physical Theory from Plato to Galileo (1908)   131

back cover should read:
PETER BARKER is Professor of the History of Science, University of
Oklahoma, Norman.

A and CMW

Introduction

Pierre Duhem: Life and Works

Pierre Maurice Marie Duhem was born in Paris on June 9, 1861. His father, Pierre-Joseph Duhem, was of Belgian origin, the oldest of eight children residing in the northern industrial city of Roubaix. After the death of his parents, Pierre-Joseph Duhem abandoned his studies to provide for the family, but it is said that, later in life, he was always seen with the work of a Roman author under his arm. Pierre Duhem's mother, Marie Fabre, was descended on her mother's side from the Hubault-Delormes, who had settled in Paris during the seventeenth century. Her father's family had originally come from the southern town of Cabesprine, near Carcasonne, and it was there, in a house they still maintained, that Pierre Duhem died on September 14, 1916.

Duhem was well educated. Starting at the age of seven, he was given private lessons in a group of four students, in grammar, arithmetic, Latin, and catechism. Those were difficult years in Paris, with the Franco-Prussian War raging until the armistice in February 1871 and the Paris Commune following in March. The fall of 1872 brought personal tragedies to the Duhem family: A diphtheria epidemic killed Duhem's younger sister Antoinette and his recently born brother Jean, leaving only Pierre and Antoinette's twin sister Marie. Duhem continued his education as a *demi-pensionnaire* (or external student) at a Catholic boarding school, the Collège Stanislas in Paris, in 1872 and for the next ten years. In 1882 he entered a prestigious secular institution of higher education, the Ecole Normale Supérieure. He was first in his class when he entered the Ecole

Normale, and he remained first throughout his five years there. He received a *license* in mathematics and another in physics at the end of the academic year 1883–1884. In his final year, Duhem was offered a position in Louis Pasteur's laboratory as a chemist-bacteriologist, though he refused it because of his desire to work in theoretical physics.

One of the turning points in Duhem's career occurred during the academic year 1884–1885. Duhem presented a thesis in physics for his doctorate. The thesis, *Le Potentiel thermodynamique et ses applications à la mécanique chimique et à l'étude des phénomènes électriques*, was rejected by a panel of three scholars. The speculation was that the panel, chaired by Gabriel Lipmann, made a political decision. A version of the thesis was published the following year by the prestigious French scientific publisher Hermann. Duhem defended another thesis in applied mathematics, *Sur l'aimantation par influence*, and received his doctorate in October 1888.

It would be difficult to fully understand these events without delving deeply into the social, cultural, and intellectual context of France at the end of the nineteenth century. At a time when French scientists were predominantly liberal and antireligious, Duhem's conservative political and religious views were certainly significant factors. The structure of the French scientific community was also surely a force in the affair. The specific motives generally cited in the case, however, were Lipmann's "jealousy" and the fact that Duhem's thesis refuted the cherished theses of Marcelin Berthelot, a friend of Lipmann, and a power in the French scientific establishment. It was reported that Berthelot had said: "This young man will never teach in Paris." Berthelot's edict came true. Duhem spent his academic career in provincial universities far from Paris, the center of academic life in France. His teaching positions brought him from Lille to Rennes and to Bordeaux, but not to Paris.

Duhem assumed the position of Maître de Conférences at the Faculté des Sciences at Lille in October 1887. There he met Adèle Chayet, whom he married in October 1890. Their daughter, Hélène, was born in September 1891. Tragically, Adèle died in childbirth the following summer. The newborn child also did not survive. Duhem did not remarry. He left the upbringing of Hélène to his mother, who lived with him after Pierre-Joseph died. The situation in Lille soured for Duhem. Never one to back off from a dispute, he fought with the dean of his faculty over a minor misunderstanding. The misunderstanding escalated to enormous proportions. Duhem requested and received a change of position at the end of the academic year 1893.

During these formative years, Duhem worked very hard on his science. He published six books: a two-volume work on hydrodynamics, elasticity, and acoustics; his lectures on electricity and magnetism, in three volumes; and an introduction to physical chemistry. He also began a series of articles describing his philosophy of science (see chapters 1 and 2 of this volume).

In October 1893, Duhem left Lille for Rennes. He remained there for only one year, leaving for Bordeaux in October 1894. Duhem was hoping for a position in Paris. His friends advised him to accept the position in Bordeaux, saying improbably: "The road to Paris goes through Bordeaux." But the road to Paris got longer and longer. Duhem remained in Bordeaux until the end of his life, a little more than twenty years later.

The prodigious quantity and quality of Duhem's publications in many fields of science, the philosophy of science, and the history of science did not change his situation. Very late in life, he was approached about the newly created chair in the History of Science at the Collège de France, but he refused to be a candidate for it. The proud and stubborn Duhem told his daughter: "I am a theoretical physicist. Either I will teach theoretical physics at Paris or else I will not go there."

Duhem's productivity during the Bordeaux years was incredible. His *curriculum vitae*, written in 1914 on the occasion of his nomination to the Académie des Sciences, lists more than three hundred fifty items, about fifty of which were books, including such masterpieces as *The Aim and Structure of Physical Theory* and *To Save the Phenomena* (see chapter 6 of this volume). The main difficulty in interpreting Duhem's work is not its quantity, but Duhem's habit of using earlier work in new contexts, sometimes changing it in subtle ways. This practice has passed unnoticed by many earlier readers and is one of the main motives for presenting our new translations.

Duhem's early views reflected late-nineteenth-century positivism. Physical theory was no more than an aid to memory,[1] summarizing and classifying facts by providing a symbolic representation of them, and quite different from common sense and metaphysics, especially the mechanical theories fashionable at the time (see chapter 1). Duhem's position was immediately attacked by a Catholic engineer, Eugène Vicaire, on the ground that separating physics from metaphysics implied that physics was the only real knowledge (another positivistic thesis) and thus conceded too much to skepticism. Vicaire had raised an important point for turn-of-the-

---

1. For an exposition and criticism of this positivistic view, see Duhem's review of Mach's *Mechanics*, chapter 5.

century Catholics, because the Church was officially committed to neo-Thomism, with its generally rationalist apologetics.[2] In reply, Duhem adopted a quasi-Thomist position: Metaphysics is a real form of knowledge, more excellent than physics but separated from it in that it has different objects and is governed by different methods (see chapter 2). This immediate response fitted well into the framework of neo-Thomism, though it did not go as far as to reunite the disparate forms of knowledge into an overall system of subaltern and subalternated sciences. Duhem's mature position was somewhat different and elaborated two key ideas: the underdetermination of theory by fact and the natural classification as the end point of physical theory (see chapters 3 and 4). The first of these is well known, but the importance of the second in Duhem's thought has not been sufficiently appreciated.

The natural classification will ultimately provide the true ontology of nature when it appears at the historical end point of physical theory. The degree to which any existing theory reflects the natural classification is not to be judged by the mind's logical faculties. It is not, for example, a question of whether a theory can be reduced to some preferred ontology. At present, we lack scientific access to the preferred ontology. Instead, the judgment is to be made by the intuitive mental faculty that Blaise Pascal had called *bon sens.* The doctrine of the natural classification, with its correlative concepts of the geometrical, or logical, mind and the intuitive mind, played an important role in Duhem's mature system, from the *Aim and Structure of Physical Theory* to *German Science* (see chapter 12).

In his own eyes, Duhem was primarily a physicist.[3] Like Ernst Mach, Wilhelm Ostwald, and others, he defended the position called energetics or energeticism, believing that generalized thermodynamics provided the foundation for all of physics and chemistry (see the conclusion of chapter 5). Duhem spent his whole scientific life advancing energetics, from his failed dissertation in physics to his mature treatise *Traité d'énergétique* (1911). Thus, Duhem's work in the history and philosophy of science can be viewed—and has been viewed[4]—as an attempt to defend the aims and methods of energetics. More recently, Niall Martin and others have

2. See R.N.D. Martin, *Pierre Duhem: Philosophy and History in the Work of a Believing Physicist* (La Salle, Ill.: Open Court, 1991), chap. 2.

3. His scientific legacy includes the Gibbs-Duhem and Duhem-Margules equations.

4. See A. Lowinger, *The Methodology of Pierre Duhem* (New York: Columbia University Press, 1941).

argued for the importance of religious motives in Duhem's work (see especially chapter 7),[5] and it has become clear in the course of Duhem's writings that he expected the end point of science, the natural classification, to harmonize with the teachings of the Catholic Church.

Whatever Duhem's initial motivation, his historical and philosophical work took on a life of its own, ranging over such diverse topics as the relations between the history of science and the philosophy of science, the nature of conceptual change, the historical structure of scientific knowledge, and the relations between science and religion. In keeping with energeticism, Duhem was anti-atomist (or anti-Cartesian; see chapter 10).

Duhem's rejection of atomism was based on his instrumentalism (or fictionalism), but it must be said immediately that instrumentalism was not the end of the matter for Duhem. He thought there *was* an ultimate truth—corresponding to the natural classification—and that some sciences (particularly thermodynamics) were steps toward it. He proposed what would today be called an instrumentalist account, not as an account of the ultimate nature of science, but as a means to combat premature claims to ultimate truth and to exclude the divisive influence of metaphysics from science.

In Duhem's account, physical theories are not ultimate explanations but representations. They do not reveal the true nature of matter but give general rules of which laws are particular cases. Theoretical propositions are not true or false but "convenient" or "inconvenient." Duhem's position on physical theories is therefore inconsistent with atomism, which refers the explanation of observable phenomena to imperceptible bodies and their motions. In the contemporary debates about light and magnetism, he rejected both James Clerk Maxwell's work, with its universal substratum, and Hendrik Lorentz's electron theory, which combined the worst features of atomism with Maxwell's position. Duhem strongly attacked the use of models by scientists such as Michael Faraday and Maxwell. He believed that the excessive use of models would lead to discontinuity with past theories and illicit claims to ontological novelty and would, by reintroducing metaphysics into physics, destroy the prevailing consensus.[6]

5. R.N.D. Martin, op. cit.

6. See Roger Ariew and Peter Barker, "Duhem on Maxwell: A Case-Study in the Interrelations of History of Science and Philosophy of Science," *Philosophy of Science Association* Volume 1 (1986): 145–156.

Duhem's rejection of atomism was coupled with a critique of inductivism, the doctrine that the only physical principles are general laws known through induction, based on the observation of facts. His critique forms a series of theses collectively known as the Duhem thesis:[7] Experiments in physics are observations of phenomena accompanied by interpretations; physicists therefore do not submit single hypotheses but whole groups of hypotheses to the control of experimentation; thus, experimental evidence alone cannot conclusively falsify isolated hypotheses. For similar reasons, Duhem rejected the possibility of crucial experiments, understood as ways of deciding between rival theories that provide parallel explanations of large classes of phenomena, on the basis of a single fact. He argued that crucial experiments resemble false dilemmas: Hypotheses in physics do not come in pairs, so crucial experiments cannot transform one of the two hypotheses at stake into a demonstrated truth.

Duhem as a philosopher weaves together two large patchworks of theses: (a) the autonomy of physics from metaphysics, entailing fictionalism or instrumentalism, the rejection of atomism and Cartesianism, and his hostility toward models; and (b) the Duhem thesis—that is, nonseparability and nonfalsifiability, entailing the critique of the Newtonian method, or anti-inductivism, and the rejection of crucial experiments. Both sets of theses are intended as historically grounded claims about the workings of science. Both sets are important in understanding Duhem's thought. The first set of theses effectively demarcates physical theory as an autonomous domain apart from other domains—that is, it rejects any external method—and the second set then operates on the internal workings of physical theory. Having set apart physical theory, Duhem asserts that no internal method leads inexorably to the truth.

Duhem's historico-philosophical works were discussed by the founders of the twentieth-century philosophy of science, including Mach, Henri Poincaré, the members of the Vienna Circle, and Karl Popper. A revival of interest in Duhem's philosophy began with W. V. Quine's reference in 1953 to the Duhem thesis. As a result, Duhem's philosophical works were translated into English as *The Aim and Structure of Physical Theory* (1954) and *To Save the Phenomena* (1969). By contrast, little of Duhem's extensive historical corpus—*Etudes sur Léonard de Vinci*, 3 vols. (1906–1913),

---

7. On the various things philosophers have attributed to Duhem under this name, and their relation to what Duhem actually said, see Roger Ariew, "The Duhem Thesis," *British Journal for the Philosophy of Science* 35 (1984): 313–325.

and *Le Système du monde*, 10 vols. (1913–1959), for example—has been translated.

In his more properly historical studies, Duhem argued that there are no abrupt discontinuities between medieval and early modern science (the so-called continuity thesis), that religion played a positive role in the development of science in the Latin West (see chapter 7), and that the history of physics could be seen as a cumulative whole, defining the direction in which progress could be expected.

It was while writing *The Origins of Statics* in 1904 that Duhem came across an unusual reference to a then-unknown medieval thinker, Jordanus de Nemore. His pursuit of this reference, and the research to which it led, is widely acknowledged to have created the field of the history of medieval science. When Duhem wrote *The Evolution of Mechanics* in 1903, he dismissed the Middle Ages as scientifically sterile. Similarly, Duhem's history of chemical combination, *Le Mixte et la combinaison chimique*, published in book form in 1902, had jumped from Aristotle's concept of *mixtio* to modern concepts. *The Origins of Statics*, however, contained a number of chapters on medieval science: One chapter treated Jordanus de Nemore; another treated his followers; a third argued their influence on Leonardo da Vinci.

In the second volume (1905–1906), Duhem greatly extended his historical scope. As expected, he covered seventeenth-century statics, but he also returned to the Middle Ages, devoting four chapters to geostatics, including the work of Albert of Saxony in the fourteenth century. *The Origin of Statics* is thus a transition from Duhem's early conventional histories to the later work for which he is best known, the three-volume *Etudes sur Léonard de Vinci* and the ten-volume *Le Système du monde*, in which his thesis of the continuity of late medieval and early modern science is fully displayed.

Duhem did not succeed in finishing *Le Système du monde*. He intended it as a twelve-volume work on the history of cosmological doctrines, ending with Copernicus. He completed nine volumes, the first five being published from 1914 to 1919 and the next four having to wait until the 1950's; a tenth, incomplete volume was also published then. After he was finished with *Le Système du monde*, he intended to write a three-hundred-page summary of his results. He did not have the time to accomplish what would have surely been a brilliant volume. Such essays as the ones he wrote for the *Catholic Encyclopaedia* on the history of physics and the summary of his main historical discoveries, with his account of his scientific career, remain the best and most concise expositions of his historical views (see chapters 8 and 11).

Unlike his philosophical work, Duhem's historical work was not sympathetically received by his influential contemporaries. The continuity of medieval and early modern science was rejected by Antonio Favaro, the editor of *Le Opere di Galileo Galilei*, as early as 1916. The next generation of historians of science, led by Alexandre Koyré and Anneliese Maier, constructed elaborate metaphysical justifications for their repudiation of continuity. Their work influenced Thomas Kuhn and others who made "scientific revolutions" a central feature of their historical accounts. We have argued elsewhere that much of this work misrepresents Duhem and that continuity may be defensible as a historiographical thesis, though perhaps not in quite Duhem's form.[8] The work of Kuhn and later historically oriented philosophers and sociologists of science, however, at least attempts to reintegrate the philosophical and historical studies that Duhem pursued together but that have been separated for a good part of the twentieth century.

In an appendix to *The Aim and Structure of Physical Theory*, amplifying the article translated here as chapter 3, Duhem makes the connections between his history and his philosophy of science very clear. For Duhem, the natural classification that will appear at the end point of science is the ideal physical theory. The natural classification will also be the only reliable source for cosmology. Earlier cosmologies will be objectionable for the same reasons as were earlier physical theories:

> It is not enough for cosmologists to know very accurately the doctrines of contemporary theoretical physics; they must also know past doctrines. In fact, current theory need not be analogous with cosmology, but with the ideal theory toward which current theory tends by a continual progress. Therefore, it is not up to philosophers to compare physics as it is now with their cosmology, by congealing science in some manner at a precise moment of its evolution, but rather to appreciate the development of theory and to surmise the goal toward which it is directed. Now, nothing can guide them safely in conjecturing the path that physics will follow if not the knowledge of the road it has already covered. If we perceive in the blink of an eye an isolated position of the ball hit by

---

8.   Roger Ariew and Peter Barker, "Duhem and Continuity in the History of Science," *Pierre Duhem, Revue Internationale de Philosophie* 46 (1992); Peter Barker and Roger Ariew, eds., *Revolution and Continuity: Essays in the History and Philosophy of Early Modern Science* (Washington, D.C.: Catholic University of America Press, 1991).

a handball player, we cannot predict the end point the player aimed at; but if our glance follows the ball from the moment the player's hand moves to strike it, our imagination, prolonging the trajectory, marks in advance the point the ball will strike. Thus, the history of physics lets us guess at a few features of the ideal theory to which scientific progress tends, that is, the natural classification that will be, as it were, an image of cosmology.[9]

As one can see, the philosophy of science and the history of science are inextricably interwoven for Duhem. Thus, we believe that it is appropriate to display Duhem's philosophical ideas and some of his historical ideas in a single volume, on the ground that they are mutually supportive. Although in our presentation the philosophical papers precede the papers on history, the reader will find that many personalities and episodes recur. (Indeed, the 1911 "History of Physics" presented here as chapter 8, may be used as a handy reference with which to find additional information about the scientists, theories, and experiments discussed elsewhere.) We have chosen to order the material chronologically, in order to preserve the historical development of Duhem's own thought. We hope this translation will stimulate new interest in Duhem's work and perhaps allow Duhem a fairer hearing than he has received so far. But we also believe that the material gathered here has independent value as an image of physics—and the study of its history and philosophy—as it appeared before the founding of the relativity and quantum theories. In addition to studying Duhem, then, our aim is to further the contextual study of the history of science.

A Note on the Translation

Duhem's prose is not ornate, but it is prolix. Readers familiar with A. A. Milne may recall the closing sentence of the story in which Christopher Robin and Pooh rescue Piglet from the flood, a sentence so long that, in the course of it, Piglet goes to sleep and nearly falls out of a window—a sentence longer than the sentence you are now reading, but not yet a sentence of average length for Duhem. The main challenge facing the translator is therefore to dissect Duhem's long French sentences into short English sentences.

9. From "Physics of a Believer," originally published in *Annales de philosophie chrétienne* I, 4th series (1905), and included as an appendix of *The Aim and Structure of Physical Theory*.

We have attempted a translation that, we hope, follows Duhem's prose more closely than some earlier efforts yet is more technically accurate in scientific matters. But where an overly literal translation would have obscured Duhem's point as we understand it, we have followed our sense of the argument in choosing our words. We have also tried to be consistent in our translation of technical terms, with small exceptions dictated by style. Duhem overuses the noun *ensemble*. We have generally translated it as "group" and occasionally as "set."

Unspecified human beings have been rendered in the plural; for example, "*le lecteur Français . . . il*" generally becomes "French readers . . . they." We have also attempted to avoid the use of the English "one" to translate the French *on*, generally preferring to substitute "we" or "they" according to context.

To enable the interested reader to follow Duhem's remarks into the current literature, we have given the accepted modern version of individuals' names, following *The Dictionary of Scientific Biography* (New York: Scribner's, 1970–1980) and Thorndike's *History of Magic and Experimental Science* (New York: Columbia University Press, 1923–1958) where possible.

Text enclosed in square brackets, particularly in footnotes, was added by the translators.

Where Duhem quoted from another writer's book and an English translation already exists, we have quoted the existing translation and cited it in a footnote. With very few exceptions, we have not corrected these translations, so that the interested reader may locate the corresponding original passages.

Acknowledgments

This project received much assistance from numerous people and institutions. It owes its inspiration to the many Duhem scholars who have toiled mightily to enable others to appreciate Duhem's accomplishment for its own sake, such as Anastasios Brenner, Paul Brouzeng, Michael Crowe, Roberto Maiocchi, Niall Martin, Stanley Jaki, and William Wallace. We cannot thank these scholars enough. Moreover, Daniel Garber and Bernard R. Goldstein showed incredible patience while we were playing hooky from our long-term collaborative projects with them, on Descartes and Kepler, respectively. We wish to acknowledge Virginia Polytechnic Institute and State University (Virginia Tech), whose Department of Philosophy and Center for the Study of Science in Society supported the project in numerous ways. We especially thank Marjorie Grene, who graciously enabled a French schoolboy and an English schoolboy to achieve American prose occasionally. Finally, special recognition is due our spouses, Susan Ariew and Catherine Webb, who read, transcribed, inputted, and listened to various drafts of the manuscript, and who at times even released us from our familial duties to work on the volume (however pleasant playing with David, Daniel, Karen, and Erika might have been).

Selected Bibliography of Duhem's Works

1886, *Le Potentiel thermodynamique*, Paris: Hermann.

1892, "Quelques réflexions au sujet des théories physiques," *Revue des questions scientifiques* 31: 139–177 (chapter 1 of this volume).

1892, "Notation atomique et hypothèses atomistiques," *Revue des questions scientifiques* 31: 391–454.

1893, "Physique et métaphysique," *Revue des questions scientifiques* 34: 55–83 (chapter 2 of this volume).

1893, "L'École anglaise et les théories physiques," *Revue des questions scientifiques* 34: 345–378 (chapter 3 of this volume).

1894, "Quelques réflexions au sujet de la physique expérimentale," *Revue des questions scientifiques* 36: 179–229 (chapter 4 of this volume).

1900, *Les Théories électriques de J. Clerk Maxwell: Étude historique et critique*, Paris: Hermann.

1902, *Le Mixte et la combinaison chimique*, Paris: C. Naud. (2nd ed., Paris: Fayard, 1985.)

1902, *Thermodynamique et chimie*, Paris: Gauthier-Villars. (English trans., George K. Burgess, *Thermodynamics and Chemistry*, New York: Wiley, 1903.)

1903, *L'Évolution de la mécanique*, Paris: A. Joanin. (English trans., Michael Cole, *The Evolution of Mechanics*, Alphen aan den Rijn: Sijthooff and Noordhoff, 1980.)

1903, "Analyse de l'ouvrage de Ernst Mach," *Bulletin des sciences mathematiques* 27: 261–283 (chapter 5 of this volume).

1905–1906, *Les origines de la statique*, 2 vols., Paris: Hermann. (English trans., Grant F. Leneaux, Victor N. Vagliente, and Guy H. Wagener, *The Origins of Statics*, Dordrecht: Kluwer, 1991.)

1906, *La Théorie physique, son objet et sa structure*, Paris: Chevalier et Rivière. 2nd ed. 1913. (English trans., Phillip Wiener, *The Aim and Structure of Physical Theory*, Princeton: Princeton University Press, 1954.)

1908, *SOZEIN TA PHAINOMENA, essai sur la notion de théorie physique de Platon à Galilée*, Paris: Hermann. (English trans., Edmond Dolan and Chaninah

Maschler, *To Save the Phenomena*, Chicago: University of Chicago Press, 1969; chapter 7 and conclusion are chapter 6 in this volume.)

1906–1913, *Études sur Léonard de Vinci*, 3 vols. Paris: Hermann.

1911, "History of Physics," *Catholic Encyclopaedia*, New York: R. Appleton, pp. 47–67 (chapter 8 of this volume).

1911, *Traité d'énergétique ou de thermodynamique générale*, 2 vols., Paris: Gauthier-Villars.

1912, "La Nature du raisonnement mathématique," *Revue de philosophie* 12: 531–543 (chapter 9 of this volume).

1913–1959, *Le Système du monde, histoire des doctrines cosmologiques de Platon à Copernic*, 10 vols., Paris: Hermann.

1915, *La Science allemande*, Paris: Hermann. (English trans., John Lyon, *German Science*, La Salle, Ill.: Open Court, 1991; "Quelques réflexions sur la science allemande" from the book is chapter 12 in this volume.)

1917, "Notice sur les travaux scientifiques de Duhem," *Mémoires de la société des sciences physiques et naturelles de Bordeaux* 7/:71–169 (parts II and III are chapters 10 and 11 in this volume.)

1985, *Medieval Cosmology*, trans. Roger Ariew, Chicago: University of Chicago Press (partial translation of vols. 7–9 of *Le Système du monde*).

1994, *Lettres de Pierre Duhem à sa fille Hélène*, ed. Stanley L. Jaki, Paris: Beauchesne.

Selected Bibliography of Works on Duhem

Ariew, Roger. 1984. "The Duhem Thesis," *British Journal for the Philosophy of Science* 35: 313–325.

Ariew, Roger, and Peter Barker. 1986. "Duhem on Maxwell: A Case-Study in the Interrelations of History of Science and Philosophy of Science," *Philosophy of Science Association* 1: 145–156.

———, eds. 1990. *Pierre Duhem: Historian and Philosopher of Science*, *Synthèse* 83, no. 2, including: F. Jamil Ragep, "Duhem, the Arabs, and the History of Cosmology"; Stephen Menn, "Descartes and Some Predecessors on the Divine Conservation of Motion"; William Wallace, "Duhem and Koyré on Domingo de Soto"; Robert S. Westman, "The Duhemian Historiographical Project"; Steven J. Livesey, "Science and Theology in the Fourteenth Century: The Subalternate Sciences in Oxford Commentaries on the *Sentences*"; Roger Ariew, "Christopher Clavius and the Classification of the Sciences"; André Goddu, "The Realism That Duhem Rejected in Copernicus"; and Peter Barker, "Copernicus, the Orbs, and the Equant."

———. 1990. *Pierre Duhem: Historian and Philosopher of Science*, *Synthèse* 83, no. 3, including: Anastasios A. Brenner, "Holism a Century Ago: The Elaboration of Duhem's Thesis"; R.N.D. Martin, "Duhem and the Origins of Statics: Ramifications of the Crisis of 1903–1904"; Phillip L. Quinn, "Duhem in Different Contexts: Comments on Brenner and Martin"; Don Howard, "Einstein and Duhem"; Roberto Maiocchi, "Pierre Duhem's *The Aim and Structure of Physical Theory*: A Book Against Conventionalism"; Richard M. Burian, "Maiocchi on Duhem, Howard on Duhem and Einstein: Historiographical Comments"; Andrew Lugg, "Pierre Duhem's Conception of Natural Classification"; Ernan McMullin, "Comment: Duhem's Middle Way"; Michael J. Crowe, "Duhem and the History and Philosophy of Mathematics"; and Douglas Jesseph, "Rigorous Proof and the History of Mathematics: Comments on Crowe."

Brenner, Anastasios. 1990. *Duhem, science, réalité et apparence*, Paris: Vrin.

———, ed. 1992. *Pierre Duhem, Revue internationale de philosophie* 46, including: Maurice A. Finocchiaro, "To Save the Phenomena: Duhem on Galileo"; Alain Boyer, "Physique de croyant? Duhem et l'autonomie de la science"; Roger Ariew and Peter Barker, "Duhem and Continuity in the History of Science"; Brian S. Baigrie, "A Reappraisal of Duhem's Conception of Scientific Progress"; Angèle Kremer-Marietti, "Measurement and Principles: The Structure of Physical Theories"; Roberto Maiocchi, "Duhem et l'atomisme"; and Anastasios Brenner, "Duhem face au post-positivisme."

Brouzeng, Paul. 1987. *Duhem, science et providence*, Paris: Belin.

Harding, Sandra G., ed. 1976. *Can Theories Be Refuted? Essays on the Duhem-Quine Thesis*, Dordrecht: Reidel, including: W. V. Quine, "Two Dogmas of Empiricism"; Carl Hempel, "Empiricist Criteria of Cognitive Significance"; Adolph Grünbaum, "The Duhemian Argument"; Larry Laudan, "Grünbaum on 'The Duhemian Arguments'"; C. Giannoni, "Quine, Grünbaum, and the Duhemian Argument"; G. Wedeking, "Duhem, Quine, and Grünbaum on Falsification"; Mary Hesse, "Duhem, Quine, and a New Empiricism"; Adolph Grünbaum, "Is It *Never* Possible to Falsify a Hypothesis Irrevocably?"; and short selections from Duhem, Karl Popper, Thomas Kuhn, Imre Lakatos, and Paul Feyerabend.

Jaki, Stanley L. 1984. *Uneasy Genius: The Life and Work of Pierre Duhem*, The Hague: Martinus Nijhoff.

———. 1988. *The Physicist as Artist*, Edinburgh: Scottish Academic Press.

———. 1992. *Reluctant Heroine: The Life and Work of Hélène Duhem*, Edinburgh: Scottish Academic Press.

Jordan, Emile. 1917. "Duhem, Pierre, mémoires," *Société des sciences physiques et naturelles de Bordeaux* 7/I: 3–40.

Lowinger, Armand. 1941. *The Methodology of Pierre Duhem*, New York: Columbia.

Maiocchi, Roberto. 1985. *Chimica e filosofia, scienza, epistemologia, storia e religione nell'opera di Pierre Duhem*, Firenze: La Nuova Italia.

Martin, R.N.D. 1982. "Saving Duhem and Galileo," *History of Science* 25: 302–319.

———. 1991. *Pierre Duhem: Philosophy and History in the Work of a Believing Physicist,* La Salle, Ill.: Open Court.

———. 1991. "The Trouble with Authority: The Galileo Affair and One of Its Historians," *Modern Theology* 7: 269–280.

Nye, Mary Jo. 1986. *Science in the Provinces: Scientific Communities and Provincial Leadership in France, 1860–1930,* Berkeley: University of California Press.

Paul, Harry W. 1979. *The Edge of Contingency: French Catholic Reaction to Scientific Change from Darwin to Duhem,* Gainesville: University Presses of Florida.

Pierre-Duhem, Hélène. 1936. *Un Savant français, Pierre Duhem,* Paris: Plon.

1

Some Reflections on the Subject of Physical Theories[1]

This article contains the first statements of some of Duhem's characteristic the-
ses, but not the so-called Duhem Thesis. At several points in the argument,
Duhem refers to the teaching of physics. This is not accidental: The article is
based on the opening lectures for his course on mathematical physics and crys-
tallography at Lille.

1. On the Aim of Physical Theory

Placed in contact with the external world in order to understand it, the
human mind first encounters the domain of facts. It sees that a piece of
amber, rubbed by a silk rag, attracts a pith ball suspended from a silk
thread, at a distance; that a piece of glass, rubbed with a woollen rag, does
the same thing; that a piece of copper, rubbed with the same woollen rag,
also does the same thing, provided that the piece of copper and the woollen
rag are both carried by a glass sleeve, etc. Each observation, each new
experiment, presents a new fact.

The understanding of a great number of facts forms a confused mass
that constitutes, properly speaking, *empiricism*.

This understanding of particular facts is no more than the first level of
understanding of the external world. The mind arrives at the understanding of *experimental laws* through induction, transforming the facts it has
come to understand. Thus, the facts that we have just mentioned, and

1. ["Quelques réflexions au sujet des théories physiques," *Revue des questions sci-*
entifiques 31 (1892): 139–177.]

1

other similar facts it is possible to observe, lead the mind, through induction, to this law: When similarly rubbed, all bodies become capable of attracting a pith ball suspended on a silk thread. Creating a new word to express the general property that this law asserts, the mind says: Through suitable rubbing, all bodies are electrified.

It falls to philosophers to analyze the mechanism of the inductive procedure that allows us to pass from facts to laws, and to discuss the generality and certainty of the laws thus established. I have no wish to further consider, here, the examination of these questions; rather, I wish to study the very understanding of facts.

The understanding of experimental laws constitutes *purely experimental science* and is elevated above empiricism as a law is above particular fact.

But purely experimental science is not the final stage in understanding the external world. Above it is *theoretical science*. What we propose to study is the nature of this science, taking as an example the theory closest to perfection, the one that has received the name *mathematical physics*.

Theoretical science has as its aim to relieve the memory and to assist it in retaining more easily the multitude of experimental laws. When a theory is constructed, the physicist, instead of having to retain (in memory) a multitude of isolated laws, only has to store in memory a small number of definitions and propositions stated in the language of mathematics. The consequences that analysis permits the physicist to logically deduce from these propositions have no *natural* connection with the laws that form the proper object of his studies. But they provide him with an image. This image is more or less representative. But when the theory is good, this image suffices to replace the understanding of experimental law in applications that the physicist wishes to make.

Let us explain all this by analyzing how a physical theory is constructed.

2. On Definitions in Physical Theory

In the first place, a physicist who wishes to found a theory that will bring together a collection of laws takes the diverse physical concepts on which these laws bear, one after another. An algebraic or geometric magnitude with properties that represent its most immediate properties is made to correspond to each physical concept.

Thus, how do we go about constructing a theory of heat? The most elementary laws that we are trying to coordinate through this theory bring into play a concept—that of *warmth*. This concept presents certain immediate features: For example, we understand that two bodies of the same nature or different in nature might be as warm as each other or that one of the two might be more warm or less warm than the other. We understand that two parts of the same body may or may not be as warm as each other. We know that if body A is warmer than body B and body B is warmer than body C, body A is warmer than body C.

These features, which are essential to the concept of *warmth*, do not permit the *measurement* of the object of this concept—that is, to regard it as a *magnitude*.

In fact, for an object to be measurable, it is necessary that the concept we have of this object present not only all the features that we have just enumerated but also the feature of *additivity*. But we have not conceptualized warmth as susceptible to addition. We understand well enough what these phrases mean: Body A is as warm as body B; body A is warmer than body B. But we do not understand what statements such as these would mean: The *warmth* of body A is equal to the *warmth* of body B plus the *warmth* of body C; body A is seventeen times warmer than body B or is three times less warm than body B.

Thus, warmth is not conceptualized by us as susceptible of addition. For us, this concept is not reducible to a magnitude.

But if the concept of warmth is not reducible to a magnitude, this does not in any way prevent physicists from making a certain magnitude correspond to it. They call this magnitude *temperature*. They choose this magnitude in such a manner that its most simple mathematical properties *represent* the properties of the concept of warmth.

Thus, warmth is presented as a proper feature of each of the points of a body. We conceptualize each point of a body as being equally warm, less warm, or warmer, than every other point. To each point of a body we make correspond a definite value of temperature. The concept of warmth does not imply any concept of direction. We would not understand what was meant by this phrase: At the point M in a body it gets warmer following the direction MN than following the direction MN'. Temperature will thus be a simple algebraic quantity and not a geometric magnitude.

We make two equal values of temperature correspond to two points that are as warm as each other. We make two unequal values of temperature correspond to two points that are not equally warm, and in such a manner that the higher value of temperature corresponds to the warmer point.

This operation establishes a correspondence between the concept of warmth and the algebraic magnitude that we call temperature. There is no sort of *natural* relationship between these two ideas: warmth and temperature. Warmth is pleasant or unpleasant to us; it warms us and it burns us. A temperature may be added to another temperature; it may be multiplied or divided by a number.

But, by virtue of the correspondence established between these two ideas, the one becomes the *symbol* of the other to such an extent that telling me that the temperature of a body has a definite value tells me which bodies are as warm as or less warm or warmer than this body.

By virtue of this correspondence, all physical laws relative to warmth and stated by propositions of ordinary language are symbolically translated by mathematical propositions concerning temperature.

Thus, instead of saying that all the points of the body are as warm as one another, we will say that the temperature has the same value at all points in the body.

Instead of saying that body A is warmer than body B, we will say that the temperature of body A has a greater value than the temperature of body B.

The example that we have just developed makes the general characteristics presented by the definition of a physical quantity quite evident. For what we have just said about temperature could be repeated—at least in its essentials—about all the definitions of the magnitudes that we find at the beginning of any physical theory whatsoever. We thus see that physical definitions construct what may truly be called a vocabulary: Just as a French dictionary is a collection of conventions making a name correspond to each object, so, in a physical theory, the definitions are a group of conventions making a magnitude correspond to each physical concept.

Among the essential features that such a definition presents, there is one that we wish above all to make clear: It is that such a definition is to a high degree arbitrary. Although in geometry we can only have a single good definition of a given concept—of a right angle, for example—in physics we may have an infinity of definitions of a concept, for example, of the concept of temperature or the concept of light intensity.

The physical concept that we are concerned to represent possesses a certain number of fundamental properties. The magnitude intended to symbolize it must present a certain number of essential features to represent these properties. But any magnitude which presents these features may be taken for a symbol of the physical concept that concerns us.

Thus, temperature may present the following characteristics:

It has the same value for two equally warm bodies.

It has a greater value for body A than for body B if body A is warmer than body B.

But any magnitude that presents these two features may be taken as *temperature*. The other properties that serve to complete this definition do not matter much. It does not matter that temperature might be defined relative to volumes, pressures, electromotive forces, and so on.

3. On Hypotheses in Physical Theory

The definition of the various magnitudes capable of symbolizing the concepts of a theory constitutes the first of the operations from which that theory will arise. Let us see what series of operations may then develop and realize the theory.

Between the various magnitudes that we now suppose to be defined, we will establish a certain number of relations, expressed by mathematical propositions. We will name these relations *hypotheses*.

Taking these hypotheses as principles, we develop their logical consequences from them.

By virtue of the stated definitions, there are some among these consequences that may be translated into propositions bearing uniquely on physical concepts; that is, propositions presenting the form of experimental laws. These consequences are given the name *experimentally verifiable consequences* of the theory.

These experimentally verifiable consequences of the theory form two classes: consequences translated by an exact experimental law and consequences with translations that contradict an experimental law.

If the experimentally confirmed consequences of the theory form an extended and varied group, the theory will have fulfilled the aim that was assigned to it. It will allow physicists to forget all the experimental laws that they can recover by means of it, remembering only a few definitions and a few hypotheses. Such a theory will be a good one.

If, on the other hand, the theory furnishes only a small number of consequences verifiable by experiment, it will not have fulfilled its goal of coordination. It will not be a good one.

All this is easy to understand. It is unnecessary to say more about it. But there is one point, as delicate as it is important, to which it is necessary to return: the choice of hypotheses. How shall we express these propositions

that are intended to serve as principles for the theory? By what rules shall
we choose them?

In principle, we are absolutely free to make this choice as we see fit. To
the extent that consequences logically deduced from these hypotheses by
mathematical analysis provide us with a symbol of a great number of exact
experimental laws, no one has the right to ask us to give an account of the
considerations that dictated this choice to us.

This is what Nicholas Copernicus expressed so well at the beginning of
his book *De Revolutionibus*, in saying: "It is neither necessary that these
hypotheses be true, nor even that they be likely, but one thing is sufficient;
namely, that the calculation to which they lead agrees with observations."[2]

But in fact, it is quite certain that this choice is not made by chance.
There are general methods for establishing the fundamental hypotheses of
most theories, and classifying these methods classifies the theories at the
same time.

The ideal and perfect method would consist in accepting no hypotheses
except the symbolic translation, in mathematical language, of some of the
experimental laws from the group we wished to represent. Under these
conditions, the development of the theory in its entirety would itself be a
symbolic translation in mathematical language of reasoning capable of
being formulated in ordinary language. This reasoning would take as principles experimental laws that symbolized its hypotheses. It would have for
conclusions experimental laws that symbolize the consequences of the theory. Mathematical analysis would play no other role than to abbreviate and
assist ordinary language. All the consequences of the theory would present
the same degree of certainty and accuracy that the experimental laws taken
for hypotheses do. The experimental laws that appeared as consequences
of the theory would truly be a logical consequence of the experimental laws
taken for hypotheses.

Such a theory would offer absolutely nothing *hypothetical*. Its author
would justly be able to pronounce the famous *hypotheses non fingo* of Newton.

Let it be said immediately: Physics presents to us several theories that
approach this ideal more or less. It does not offer us any theories that realize it completely. Newton may put forward *hypotheses non fingo*. Ampère

2. *Neque enim necesse est eas hypotheses esse veras; imo, ne versimiles quidem; sed sufficit hoc unum, si calculum observationibus congruentem exhibeant.* [This is not Copernicus, but Osiander in his notorious unsigned preface to *De Revolutionibus*, and the
views he is expressing are not those of Copernicus. See chapter 8, section XIII.]

may entitle his book *Théorie mathématique des phénomènes électrody-namiques uniquement déduite de l'éxperience* (*Mathematical Theory of Elec-trodynamic Phenomena Uniquely Deduced from Experiment*). In fact, it is easy to show that their hypotheses are not simple symbolic translations of experimental laws.

In this course,[3] we will take up the theory of Ampère again. We will have occasion to study in detail the hypotheses on which it rests. Let us leave it aside for the moment and take up the theory of universal attraction.

What are the experimental laws on which it rests? The laws of Kepler. What is the exact translation of these laws into a symbolic language that creates the definitions of rational mechanics? "The sun exerts on each planet an attractive force in inverse ratio to the square of the distance from the sun to the planet. The forces exerted by the sun on different planets vary as the masses of these planets. The planets exert no force on the sun." Is this the fundamental hypothesis on which Newton's theory rests? Not at all. Newton corrects the proposition that we have just stated. Then he joins to it a new proposition that is not verifiable through experiment. Then he generalizes the result obtained.

As we were saying, Newton corrects the preceding proposition: Instead of holding that the planets exert no force on the sun, as in the case of Kepler's laws, Newton states that every planet exerts a force on the sun which is equal and directly opposite to the one it receives from the sun.

Is Newton satisfied with this correction? No. He adds a proposition that is not provided by experience: If the sun was replaced by another body, the forces exerted on the different planets would be multiplied by the ratio of the mass of this new body to the mass of the sun.

Is this all? Not yet. Newton generalizes the result obtained, and it is only through this generalization that he is able to state the fundamental principle of his theory: Two material bodies, the dimensions of which are negligible in comparison to their separation, are subject to a mutual attrac-tion proportional to the product of the masses of the two bodies and in inverse ratio to the square of the distance that separates them.

What has Newton done, then? Has he taken as a hypothesis the sym-bolic translation of one or several experimental laws? Not at all. The exper-imental laws placed at the beginning of his theory are only the particular

3. [Duhem is referring to the *Course on Mathematical Physics and Crystallogra-phy*, in the Faculty of Sciences at Lille, of which this forms part. In the present paper he does not, as a matter of fact, return to Ampère.]

consequences, considered exactly or simply approximately, of a particular proposition. He has taken this proposition as a hypothesis.

This is the general procedure employed by all theoreticians. In order to formulate their hypotheses, they must choose from some of the experimental laws in the group that must be covered by their theory. Then, by way of correction, generalization, or analogy, they compose a proposition having these laws as exact or simply approximate consequences, and it is this proposition that they take as a hypothesis.

All theories rest on hypotheses that are not adequate translations of experimental laws. They are the results of a more or less considerable elaboration applied to these laws. Hence, we understand that all the intervening possibilities between the extremes may exist. At one extreme is the hypothesis that almost directly symbolizes an experimental law. This is the hypothesis closest to the ideal we spoke about a moment ago. The other extreme is a hypothesis so removed from experiment that its symbolic meaning is almost completely concealed and almost all physical meaning has been lost.

4. On the Limits of Theory and on the Modifications It Can Undergo

If all the hypotheses that a theory rests on were simply the symbolic translations of experimental laws, all the consequences of the theory would be translatable into laws with the same degree of certainty and accuracy as the degree of certainty and accuracy of the laws taken as hypotheses. But, as we have already said, the hypotheses on which a theory rests are never the exact translations of experimental laws. All result from a more or less profound modification imposed on experimental laws by the mind of the theorist.

Now, from the fact that the hypotheses on which a theory rests include something not in the experimental laws that suggested them, it follows that the certainty and accuracy of these laws do not entirely reappear in the consequences of the theory. The physical laws that symbolize the consequences of the theory may not be exact. You can be sure that however extensive and certain the physical theory may be, when it is pushed sufficiently far, it always leads to consequences contrary to experiment.

We have already said this, but the assertion is so important that it is worth the trouble of repeating it: A good theory is not a theory with no consequences that disagree with experiment. To take this standard would mean that there would never be any good theory. It is even probable that the creation of a good theory would surpass the abilities of the human mind. A good theory is one that symbolizes an extended group of physical laws in a sufficiently accurate manner and meets no contradictions in experiment except when we try to apply it outside the domain where we intended to make use of it.

Thus, it follows that the value of a theory has an entirely relative character. It depends on the group of laws that the theory must be used to systematically classify. A theory such as Poisson's may be good for classifying the laws of the distribution of electricity on the bodies of homogeneous conductors, but it ceases to be a good theory if we wish to use it to classify laws relating to all conducting bodies, whether homogeneous or heterogeneous. The same holds if we wish to understand the laws of the distribution of electricity on conducting bodies and dielectric bodies in a single system.

The value of a theory does not depend solely on the group of laws that we hope to summarize by means of this theory. It depends, again, on the degree of precision of the experimental methods that serve to establish or to apply these laws. In fact, we do not require that a consequence of the theory translate a physical law formally identical to the experimental law that we intend to represent. We only demand of it that it translate a physical law whose recorded deviations from this experimental law are smaller than the limit of observational error. It is, in fact, a principle that we must never forget: In physics, if the discrepancies between two laws that are different in form cannot be established by the methods of observation at our disposal, these laws must be regarded as identical.

Hence, certain consequences of a theory might be regarded as conforming to experimental laws by a physicist who deploys a given means of observation, and as contradicting experimental laws by another physicist who deploys more perfect means of observation and is capable of recognizing deviations that escape the instruments of the former.

The classical theory of gases, for example, was good for physicists while their instruments offered the same degree of precision as those of Gay-Lussac. When the inventive genius of Regnault had bestowed upon science much more subtle procedures, this theory became a bad one.

There is more: The old theory of gases, a bad theory for a physicist whose researches lay claim to all the precision required today, may remain good for an engineer or a chemist whose researches do not lay claim to an accuracy greater than we were content with in the time of Gay-Lussac.

Thus, a theory cannot be judged if we do not take into consideration the limits of the field it is intended to apply to and the degree of experimental accuracy that it supposes. Within the limits of the field in which the theory appears to be valuable, if one among its consequences deviates from an experimental law sufficiently for the deviation to be recognizable by the methods of observation that the theory accepts as controlling it, the theory must be condemned. Otherwise it must be approved.

What we have just said shows that one may, without contradiction, consider a theory good and propose to replace it with a better theory. The first theory represented a given group of experimental laws with a given approximation. The new theory will represent a more extensive group of laws or will represent the same laws with greater accuracy.

In order to replace one theory with a better theory, it is not always—indeed it is almost never—necessary to destroy the first entirely. Very often it is enough to construct a more complete theory in which the definitions and hypotheses of the first theory are recovered in their entirety but in which new definitions are introduced and new hypotheses formulated. It is in this way that, after having treated the theory of the distribution of electricity among systems containing only conducting bodies, without losing anything from that theory, we may complete it in a manner that also covers the laws of distribution on systems made up of conducting bodies and dielectrics at the same time.

Sometimes, a theory cannot be replaced with a better one except by means of deeper transformations which alter the meanings and the hypotheses on which the first theory rested. It is easy to understand how such transformations are possible.

The definition of a physical magnitude always implies a high degree of arbitrariness. This magnitude must present a generally limited number of features imposed on it by the very concept that it must symbolize. But any magnitude that presents these features is appropriate to symbolize this concept. To that extent, to represent the same concept one might in general make use of many extremely different magnitudes.

The simple changing of definitions would lead to changing hypotheses. If its concepts are represented by different magnitudes, the same experimental law will be symbolized by two different mathematical statements. But this purely formal modification may not be regarded as a true transformation of the hypothesis. It is simply a translation of the same hypothesis through the medium of different symbols, and these two statements of the same hypothesis in two systems of different symbols do not constitute two different hypotheses any more than statements of the same proposition in French, Latin, and Greek constitute three different propositions.

A hypothesis may be modified in a manner that affects its meaning more deeply.

If a hypothesis were simply the symbolic translation of an experimental law, it could not be modified except in the manner we have just indicated—at least to the extent that the experimental law continued to be considered accurate. But in reality, as we have said, all hypotheses are something other than simple translations of an experimental law. They are all the results of a transformation imposed on an experimental law by the mind of a physicist. And that is how they are modifiable. Two different physicists may subject the same experimental law to different transformations. Consequently, they will state two different hypotheses, construct two different theories, and lead to different consequences.

Thus, the closer the hypotheses on which a theory rests are to the ideal form that is the simple symbolic translation of an experimental law, the more difficult it will be to modify them. Further, in consequence, the theory will have a chance of lasting as long as the experimental laws that it represents, and being modified only by way of extension and growth, without being either altered or destroyed. On the other hand, the more hypotheses are separated from the experimental laws that led to their conception, the more physicists have put themselves into the laws' statements, and the more shaky and subject to demolition the theory will be. Thus, from now on, the purely logical considerations that we have just developed indicate to us the direction in which theoreticians must direct their efforts if they wish to bring to light a viable work.

5. On Mechanical Theories

Regrettably, it is not true that the efforts of theoreticians have always been directed in the sense that we have just indicated. For many among them, their ideal has long been and still is today very different from the one toward which we believe it is necessary to steer. To this erroneous tendency, we must attribute the incessant upheavals that theoretical physics has suffered and, consequently, the discredit into which this science has fallen in the mind of many physicists.

This false ideal is *mechanical theory*.

Let us try, first, to give a precise account of the nature of what is called a mechanical theory.

We have seen that for each physical concept, a theory must substitute a certain magnitude in the form of a symbol. This magnitude is constrained to present certain properties, which are an immediate translation of the features of the concept that it symbolizes. But except for these features, which are generally few in number, the definition of the concept remains absolutely arbitrary. In a mechanical theory, the added condition is imposed that all physical magnitudes are composed by means of geometrical and mechanical elements of a certain fictional system, where these magnitudes appear in the laws that we are going to have to relate among themselves. And all hypotheses are subject to the condition that they are the statements of the system's dynamic properties.

Let us take as an example the theory of light. There we will find certain concepts, including "color" and "intensity of monochromatic light." These concepts present a certain number of features which the magnitudes that symbolize them must reproduce in any theory. Color, for example, must be symbolized by a magnitude having a determinate value for each color and different values for different colors. Intensity must be represented by a magnitude that is always positive, having the same value at two points of equal illumination and a greater value at point A than at point B, if point A is more strongly illuminated than point B. The experimental laws of the propagation of light, of interference, reflection, refraction, and dispersion, all laws generalized as needed, will be translated by a series of hypotheses relating these different magnitudes among themselves. The set of these hypotheses forms the point of departure for a *physical theory* of light.

It is not, therefore, through a simple generalization of experimental laws that we would obtain our hypotheses if we wished to create a *mechanical theory* of light. We would accept that all of the physical concepts we meet in studying the phenomena of light must be represented by mechanical properties of a certain medium: the aether. We would try to imagine the constitution of this medium in such a way that its mechanical properties would be able to form a symbol of all the laws of optics. Color would then be symbolized by the period of a certain vibratory motion propagated in this medium, intensity by the mean kinetic energy of this motion; and the laws of propagation of light, reflection, and refraction would result from the application to this medium of theorems provided by the theory of elasticity. This is how the classical theory of light is constructed.

Many physicists desire no theory other than a mechanical theory. With Huygens, they think that in this way they are "within the true Philosophy, in which one understands the cause of all natural effects through mechanical reasons. This is what we must do, in my opinion, or else renounce all hope of ever understanding anything in Physics."

Such physicists demand that every physical magnitude be composed only of magnitudes that define the mechanical properties of a particular material system.

But their demands do not always stop there. Other requirements that vary from school to school generally come to be grafted onto the earlier ones. For some, material systems must be constructed from continuous media; for others, from isolated atoms. The first admit attractive or repulsive forces between different material elements; the others reject the existence of such forces and want material atoms to be able to act solely by contact according to the laws of collision.

Thus, when we are proposing simply to construct a physical theory, the only conditions imposed on the magnitudes we define and the hypotheses we state are from experimental laws, on the one hand, and from the rules of algebra and geometry, on the other. When we propose to construct a mechanical theory, we impose in addition the obligation to admit nothing in these definitions and hypotheses but a very restricted number of concepts of a definite nature.

The first inconvenience of such a method is that in restricting the number of elements that may be used in constructing the representation of a group of laws, physicists are left with no other resource than to complicate the combinations they make with these elements in order to respond to all of the demands of experimentation.

Let us imagine that two artists are asked to represent the form of the same object. One is allowed the use of all the resources provided by the arts of drawing. The other is permitted to make only a pencil sketch. Through the play of shadows, the first would be able to give a representation of the object in a single drawing that the second would equal with great difficulty by drawing a great number of profiles. The first artist is the image of the physicist who composes a physical theory; the second, of a physicist who constructs a mechanical theory. Examine the complexity of the media imagined by Sir W. Thomson to give an account of the laws of optics, or by Maxwell to represent electrical phenomena, and the fairness of this comparison will be understood.

The method that rejects all nonmechanical theories leads to great complications. It is also quite possible that it leads to impossibilities. Who assures us that all physical concepts and experimental laws may be symbolized by even a very complicated combination of purely mechanical concepts? Take the artist that you have forbidden to use any procedure except pencil sketching and ask for a rendering of an object's color that is obvious to everyone: It cannot be done. Is it not for an analogous reason that the most complex mechanical theories have not been able, up to now, to give a satisfactory account of Carnot's principle?

Thus, far from mechanical theory appearing to us as an ideal theory, we regard it as a theory hampered by obstacles that impose an insufficiently developed form on it and perhaps even make its development impossible. We have seen that a theory offers all the more guarantees that it is exact and that it will last to the extent that the hypotheses on which it rests are closer to simple translations of experimental laws. But among the hypotheses on which a mechanical theory rests, there are a large number that do not arise from experimentation and that flow solely from restrictive conventions arbitrarily imposed by the physicist. These latter hypotheses are the germs that kill all mechanical theories.

In fact, mechanical theories disappear from science one after another.

When we compare the consequences of a mechanical theory with experimental laws, we find some consequences verified and some consequences contradicted. When we ascend from these consequences to the hypotheses on which the theory rests, we find almost invariably that the verified consequences derive from those among the hypotheses that simply translate experimental laws, whereas the contradicted consequences derive from those hypotheses that impose the mechanical nature of the theory. Thus, physicists are led, little by little, to suppress these latter hypotheses in order to preserve only the former and to transform a mechanical theory into a physical theory. For example, this is how the branch of science that was presented for a long time as the *mechanical theory of heat* gradually became, under the name *thermodynamics*, one of the most perfect *physical* theories.

6. Theoretical Physics Is Not a Metaphysical Explanation of the Material World

If a mechanical theory, far from being an ideal theory, seems almost the furthest theory from the ideal, how do we explain the fashion that makes it the last word in science for so many physicists? Here we touch the nerve center of all the erroneous doctrines of theoretical physics.

We have sought to precisely delimit the nature and aim of theoretical physics. It is, as we have said, a system—a symbolic construction—designed to summarize in a small number of definitions and principles a set of experimental laws. This is its role, useful but modest. It is all too easy to exaggerate it.

An invincible urge pushes us to seek the nature of the material things that surround us and the basis for the laws that govern the phenomena we observe. This urge covers all human beings from the most superstitious savage to the most curious philosopher. Why would it not seize with great force those whose continuing meditations have the physical world as their object? To this urge, join the desire that all human beings naturally possess to increase the importance of an object that they have followed long and laboriously. You will easily understand how physicists are led to take the systems that they have constructed, with a view to symbolically representing experimental laws, as a metaphysical explanation of these laws.

There is more: Not only does everything within them press physicists to regard the theories they have constructed as explanations of nature, but also the people they live among exert a powerful influence on their ideas in the same direction. These people have only two ways of understanding physics: Either they demand immediate applications that satisfy their material needs or they require an explanation of the physical world that satisfies their ambition to understand everything. Thus, prudent scientists, who define the sense and limits of the laws they state with conscientious precision, are greeted with mistrust by them. But let someone present them with a more or less extensive theory as a metaphysical explanation of the universe, and they will greet these teachings with blind confidence. They will rank among the definitively established truths the views of a mind that exaggerates the importance of its conceptions to the point of falsifying their essential character. They will believe they are contemplating the very structure of the world but will have before their eyes only a fragile construction soon to be destroyed to make room for another.

Physicists are thus brought, as much by their own nature as by their environment, not to look for a systematic coordination of laws in a theory, but to look for an explanation of these laws. From then on, will their preferences be brought toward the pattern of theory that we have advocated as the ideal form or toward mechanical theory? It is easy to see that mechanical theory will appear to them as the goal toward which their efforts must lead.

In fact, let us imagine that a researcher has taken care, every time a physical magnitude has been defined, to note that this magnitude is used only to symbolize a concept that originates in experimentation, by some of its characteristics, and that in other ways its definition is entirely open. Let us imagine that great care has been taken, every time a hypothesis has been stated, to note the limited extent to which it translates an experimental law. However extensive, however fertile the theory may be, it will be quite difficult to lose sight of its exclusively symbolic character and to believe that an explanation of the laws so represented has been obtained.

On the other hand, let us imagine a researcher who has wholly invented a more or less complicated mechanism whose various properties represent a certain number of physical laws. It would be much easier to forget that, even if certain properties of the mechanism symbolize certain laws of the world, the mechanism itself does not represent the world. A complex conception has been constructed in order to represent a physical concept. We might be able to believe that, insofar as this complex conception represents the physical concept, the elements that compose this conception represent the causes that give birth to this concept in us. This error is similar to that of an engineer who, after constructing a robot, sees it imitating human movements and ends by imagining that the robot's structure represents the human organism.

An example will make this difference quite palpable.

Let a physicist introduce temperature in her theories as a magnitude devoted to symbolizing the concept of warmth and quantity of heat as a magnitude devoted to representing the weight of a certain body that a definite phenomenon is able to heat by a definite quantity. Let her introduce the principle of the equivalence of heat and work—that is, Carnot's principle—as a generalization of experimental law. However rich a harvest of consequences that the thermodynamic theory she has conceived brings her, she will certainly not take it for a metaphysical system explaining the universe.

On the other hand, let a physicist imagine a system constructed from an immense number of tiny bodies vibrating in place. Let him suppose that the mean kinetic energy of these tiny bodies is proportional to the absolute temperature. Through conveniently chosen assumptions on their number, their dimensions, the motions that animate them, and the forces that they exert on one another, he is able to deduce the principle of the equivalence of heat and work—Carnot's principle—from the application of the theorems of mechanics to these little bodies, and he will be inclined to exclaim: "This is how the world is constructed!"

Therefore, it is because many people wish to be able to say, "This is the explanation of the universe," while showing the combinations that result from the play of their own minds, that they are not in the least satisfied with a theory if it does not borrow all its elements from mechanics.

To those who want their theories to explain the nature and causes of physical laws, we oppose those who seek only a symbol of these laws in physical theory. These latter will not restrict ahead of time the number and nature of the concepts they will be allowed to combine. They will admit into their system other magnitudes besides those of geometry and mechanics. Once a quantity has been clearly defined, once the rules for handling it in reasoning and calculations and measuring it in experiments have been precisely set out, they will not deny themselves the fair use of it. If the hypotheses made using that quantity allow the satisfactory representation of the class of phenomena they study, their minds will be satisfied. They will not waste their time and effort in replacing that concept with a combination of geometrical and mechanical concepts.

Thus, in the theory of heat, they will seek to set out in a precise manner the rules for reasoning about the concepts of *temperature* and *quantity of heat*. Then, developing the chain of their deductions in conformity with these rules, they will draw consequences from them in the study of vaporization, liquefaction, dissociation, and solubility. When they see a complicated and varied multitude of phenomena clarify themselves and relate themselves to one another through the theory they have conceived, they will believe that they have achieved their aim. Let someone demand that they construct, with the aid of the concepts of space, time, and mass, complex concepts displaying analogous properties to those they attribute to temperature and quantity of heat, and they will disdain to satisfy these unreasonable requirements. Let someone reproach them, in that case, for using *occult qualities*, and they will not feel themselves affected by this criticism. They wished to clarify laws, not to reveal causes.

7. On the Role of Mechanical Theories in the History of Science

The critique to which we have just submitted the theories called "mechanical" immediately raises a difficulty: If these theories are based on an idea of the role of physics that is so completely wrong, how does it come about that they have been able to make such great progress in physics?

This objection deserves a response because it is impossible to deny the discoveries that science owes to mechanical theories. Descartes, Newton, Huygens, Laplace, Poisson, Fresnel, and Cauchy all assented to the idea that physics ought to be purely mechanical, and we owe modern physics to them. The theory of light, as it flowed from the genius of Fresnel, has been one of the most fertile theories, and it is a mechanical theory.

The objection is easy to answer.

It is always at the origin of a science that its role is most badly defined. Those who create it are inclined to exaggerate its scope more than are others. Hence, it is not at all surprising that almost all the creators of physical theory have sought to build mechanical theories. But it is not because they made use of similar theories that they made an abundant harvest of discoveries. The true situation is that, on the one hand, theories must above all show themselves fertile at the origin of theoretical physics, and on the other hand, at the origin of theoretical physics, mechanical theories must naturally be in favor. The fertility of mechanical theories during the last century and at the beginning of the present one is not, therefore, a logical consequence of the nature of these theories. There is simply a coincidence between their mechanical form, on the one hand, and the multiplicity and importance of the discoveries they have produced, on the other. This coincidence is not accidental at all but follows from the laws that preside over the development of science.

In the same way, during childhood, innocence coincides with the acquisition of an enormous mass of knowledge, without one of these features being able to be regarded as the consequence of the other. The two simply co-occur, and this happens because both derive from the laws of development of the human intellect. At the beginning of their intellectual development, children learn the most. It is also during this initial stage that children are least precisely aware of the value of their experiences.

If the opinion that we put forth here is accurate, to the extent that theoretical physics progresses, the most eminent physicists ought to understand its nature and aim better and better. Little by little, their preferences ought to abandon mechanical theories and direct themselves toward true physical theories. The latter must inherit the fertility that the former lose. Those who have closely followed the history of science up to our period cannot fail to have remarked the decline of mechanical theories and the ever-growing importance of purely physical theories.

Thus, what we have said about the nature of theoretical physics explains the changes that the methods proper to treating it have undergone in the course of history.

There is still another historical question that may perhaps be clarified by the preceding remarks.

If physicists seek an explanation of the laws of nature in their theories, they should be able to accept as satisfactory only a theory conforming to their metaphysical ideas. If philosophers believe that they find in the theories developed by physicists the physical foundation of material phenomena, they should be inspired by these theories in the construction of their metaphysical systems. A very intimate and powerful mutual influence of physics and metaphysics in each era flows from this. Cartesian metaphysics impressed its seal not only on Descartes's physics but also on Huygens's, and its essential characteristics reappear in the work of Euler and Lagrange. With Newton, a school of physics appeared with, after its founder, Laplace, Poisson, and Cauchy as the leading individuals. The history of this school, which we might call The School of Molecular Attraction, is intimately related to the history of Leibnizian ideas. In our own day, certain philosophical schools—that of Herbert Spencer, for example—are saturated by ideas borrowed from certain thermodynamic theories. This is a point that we limit ourselves to indicating here in passing, but a clear view of it illuminates the entire history of physical theory. To the extent that we are more aware of the purely symbolic role of physical theories, these theories will become more independent of fashionable doctrines, and, at the same time, we will give up the misconceived aspiration to impose our system on metaphysics. Something analogous to what took place for mathematical analysis will happen for these physical theories. Born from metaphysical and theological doctrines about the connections between the infinite and the finite—that is, between the supernatural and the natural—mathematical analysis has, in return, exercised an influence on metaphysics and theology that has not always been exempt from tyrannical aspirations. It required the genius of a Lagrange to discover, and the efforts of a century of great mathematicians to prove, that mathematical analysis had its own subject matter, with its own methods, and that it needed neither to accept the constraints of metaphysics and theology nor to impose constraints on them.

8. All the Theories of a Single Class of Phenomena Are Not Equivalent

We are not at all alone in supporting the ideas that we have just presented, and if there is an opinion that we would be happy to be able to invoke in support of our own, it is assuredly that of the illustrious analyst who wrote the following lines:

> Mathematical theories do not have as their object to reveal to us the true nature of things; that would be an unreasonable aspiration. Their unique aim is to coordinate the physical laws that experiment makes known to us, but that without the aid of mathematics we would not even be able to state.[4]

4. H. Poincaré, *Théorie mathématique de la lumière*, Préface.

The same author continues in the following terms: "The theories proposed in order to explain optical phenomena through the vibrations of an elastic medium are very numerous and equally plausible."

In these lines, we believe, a tendency appears that reigns, in our time, in all intellectual domains and is beginning to impose itself even on Mathematical Physics. This tendency consists in regarding as equivalent the different theories that can be given of a single group of laws and studying them all without preferring any among them. We would like, in a few words, to indicate how the application of this method to Theoretical Physics is illegitimate and how it is possible to avoid its use.

Assuredly, those who hold that no physical theory is an explanation of nature adequate to its object but that it is a system intended to furnish the symbol of a group of experimental laws are careful not to believe that only one theory might be able to represent a given class of phenomena. To believe otherwise would be to believe that two portraits of the same person might not differ from each other and at the same time resemble each other.

But although it is possible to make many different portraits of the same person, it does not follow that we might not reasonably prefer one of these portraits to others. Similarly, it might be the case that different theories of a single class of phenomena are logically acceptable without being equally plausible. We may have reasonable motives for preferring one among them. First, we suppose that the different theories from which we must choose are all logically acceptable, for there are theories that logic constrains us to reject or modify.

Logic leaves open the choice of hypotheses, but it requires that all the hypotheses be compatible among themselves and that they all be independent. A theory does not have the right to appeal to superfluous hypotheses. It must reduce their number to a minimum. Further, it does not have the right to gather together consequences deduced from incompatible hypotheses.

The series of deductions that follows from hypotheses and constitutes the development of the theory is subject to the laws of logic in its whole extent and in all its rigor. It is not allowed to conceal a gap there, however small it may be. If this gap can be filled, it must be. If it cannot be filled, it must, at least, be clearly delimited and marked under the form of a postulate. *A fortiori*, no contradiction may be tolerated here.

The comparison of the results of theory with facts is an operation that is not subject exclusively to the laws of deductive reasoning. Recognizing the degree of approximation that may be regarded as sufficient is somewhat arbitrary. But if we meet an experimental law that contradicts the consequences of the theory in the domain in which the theory is supposed to be applied, then the theory must be rejected or, at least, we must restrict the extent of the class of laws that it was supposed to cover.

To maintain a theory that the facts contradict shows evidence of a childlike obstinacy. As for those—and there are some—who conceal or knowingly falsify the results of experiments about the facts they are charged to observe in order to avoid ruining an idea whose success flatters their vanity, it is no longer up to logic to condemn their error but to morals or morality to permit their trickery from flourishing.

The rules that we have just stated are commonplace, or at least they ought to be—as they once were.

> The ancient theories of Physics give us complete satisfaction in this regard. All our teachers, from Laplace up to Cauchy, have proceeded in the same manner. Starting from clearly stated hypotheses, they deduced all the consequences from them with mathematical rigor, and next compared them with experiment. It seems they wished to give the same precision found in Celestial Mechanics to each of the branches of Physics.
>
> For a mind accustomed to admiring such models, it is difficult for a theory to be satisfactory. Not only will such a mind not tolerate the least appearance of contradiction, but it will require that the model's different parts be logically related one to another and that the number of distinct hypotheses be reduced to a minimum.[5]

In our time, when the rules of logic seem to be applied with difficulty, these requirements appear exaggerated to many minds, and perhaps to great minds.

Let us take an example. Maxwell wrote a treatise on electricity.[6] In this treatise he developed several different theories that are irreconcilable among themselves.[7] Some of them, like his theory of pressures in the interior of dielectrics, even contradict the best established principles of hydrostatics and elasticity. He does not concern himself with explaining these contradictions or separating the domain of each of these theories. On the contrary, he mixes and entangles them. To untangle them becomes a task so difficult that an illustrious analyst does not find it beneath his efforts. All of this work lacks precise experimental control. The facts sometimes

---

5. H. Poincaré. *Électricité et optique.* I. *Les Théories de Maxwell*, Introduction.

6. [See chapter 4, note 3.]

7. H. Poincaré, op. cit.

even show him to be wrong. Surely physicists will reject such a work? Will they not take it apart piece by piece, keeping only the good parts among its incoherences, to make them part of a more unified or logically constructed work? Not at all. Everyone admires the work of the master. Everyone reproduces it in teaching, repeating the incomprehensible things that it includes and affirming, meanwhile, with a sort of superstitious respect, that they do not understand: To listen to them, it would seem that science has the right to propose mysteries for our belief!

Let us not hesitate to reject this weakness. An illogical theory is not a mystery to which reason must bow. It is an absurdity that reason must reject without pity. What might be owed to a great physicist matters little. A powerful idea may be false. Let us admire the author and condemn the idea.

For a single class of phenomena, however, there may exist several theories, all founded on clearly stated hypotheses, all logically constructed, and all in satisfactory accord with the facts that they claim to represent. Optics offers a striking example.[8] Logically, all the theories are acceptable. Does it follow that they are all equivalent? If no logical criterion decides between them, does it follow that we cannot have any reasonable motive for preferring one to another?

Three features can serve to facilitate choosing between these different theories. They are:

> The scope of the theory.
> The number of hypotheses.
> The nature of the hypotheses.

Two theories are given. One covers a certain class of phenomena; the other covers, in a unique representation, not only that class of phenomena but also still other classes to which the mode of representation adopted by the first theory cannot be extended. Certainly, we ought to prefer the second.

Thus, the theory of reflection and refraction given by Fresnel, though valid for amorphous bodies, cannot be extended to crystals. The theory that MacCullagh and Neumann have given embraces amorphous bodies and crystals in a single presentation. The latter ought to be preferred to the former.

8. See F. E. Neumann, *Vorlesungen über die Theorie der Elasticität der festen Körper und des Lichtäthers*, and H. Poincaré, *Théorie mathématique de la lumière*.

Two theories of the same scope may invoke a different number of hypotheses: The one that invokes fewer hypotheses is certainly the better.

Last—and this is the essential point—when two theories are equal in scope and visibly invoke the same number of hypotheses, the nature of these hypotheses themselves may supply a plausible motive for choosing between them. The hypotheses on which the theories rest can be simpler, and more natural or can translate experimental results more immediately than can those on which the other theory rests.

Thus, the theory of double refraction imagined by Lamé rests on these two hypotheses:

In all directions, the medium propagates two waves.
A direction of vibration situated in the wave corresponds to each of the waves.

The meaning of these hypotheses is very clear. We see immediately which physical laws they represent. These are generalized, it is true, but not concealed. On the other hand, Cauchy's theory constructs hypotheses on the nature of the aether, whose physical sense escapes us. And these hypotheses also lack direct experimental verification. We must reasonably prefer Lamé's theory to Cauchy's.

Thus, in affirming that Mathematical Physics is not the explanation of the material world but a simple representation of the laws discovered by experimentation, we avoid the obligation to declare one theory true to the exclusion of another for each range of phenomena. We are not condemned because of that to adopt all the logically constructed theories of a single group of laws. In order to choose between them, we have very specific rules which often enough permit us to reasonably prefer one from among them to all the others.

9. On the Role that Mathematics and Experimentation Ought To Play in the Construction of Physical Theory

A physical theory is the systematic representation of a group of experimental laws. It takes for its point of departure some hypotheses chosen to represent certain of these laws. It combines them through mathematical reasoning in order to draw conclusions from them which it submits to the control of experimentation.

Experimentation therefore furnishes the material for the definitions and the hypotheses on which all theory rests. All results of theory ought to be experimental laws. Mathematical analysis is the instrument that puts the material in some form in order to draw results from it. This very simple rule fixes the relations that the mathematical method and the experimental method must maintain between themselves in the construction of a theory.

The simplest rules are often those that are violated most freely. So it is with the one we have just stated. Few respect it: Some exaggerate the role of mathematical method; others, the part of mathematical analysis.

For the first, physics must be exclusively studied by the experimental method. And they do not intend to state that indisputable truth that all physical research has experimentation for its beginning and end points. They intend to ban the use of mathematics as an instrument in the study of physics. They see it as a dangerous and fruitless instrument. It reveals nothing or demonstrates nothing except mistakes. The title of physicist and the right to teach physics is refused to those who take it up. Facts alone—facts brutal and singular—must be reported, taught, and reproduced. Every idea, considered through what makes it an idea, is false and to be condemned.

We will not delay in order to discuss a doctrine that turns the ideal physicist into a measuring instrument.

Few of those who profess this doctrine make their writings or their teaching conform to it completely. They make use of mathematics, but they want to make use of only certain branches of analysis. They find other branches too elevated and thus regard them as useless. When a definition seems too painstaking to them, or a demonstration too difficult, or a calculation too long, they declare that physics can do without it and they reject it.

How will we paint the state of confusion into which these illogical doctrines have plunged the study of natural phenomena? In order to avoid long and subtle definitions, magnitudes that have not been sufficiently defined are used all the time. In order to flee the complication of precise reasoning, or integrals that would require an exact calculus, they make do with approximations. Difficulties are concealed. Evasions are used. Sometimes veritable word games made possible by the lack of precise definitions are used to construct a theory. The mind, deceived by this sleight of hand, loses all concept of rational methods—or, at least, if it retains some concept, it abandons the theoretical study of natural phenomena with disgust. It takes refuge in works of pure observation, such as chemistry and natural history, or in researches of pure logic, such as abstract mathematics. All those who have observed the effect produced by teaching physics on the intellect of their students have been able to report this phenomenon.

The instrument of mathematics is necessary for the study of physics, and physicists must be capable of employing all the portions of that instrument whenever necessary. If a theory calls on analytical considerations that are elevated and complicated, it may be good not to present it before an audience with too little preparation. But it would be illogical to criticize the theory for the complexity of the apparatus used to construct it, unless this apparatus can be replaced with another that will be equally solid and easier to handle.

Mathematics is therefore the instrument necessary to construct all physical theory. But it is only a means, not an end. If we wish to avoid the abuse of mathematical physics, we must never lose sight of this principle.

The definitions and hypotheses that serve as points of departure for a theory ought to follow from the fundamental equations of that theory. Mathematical analysis will proceed with great care from this background equation (*mise en équation*) in making precise the conditions and restrictions to which it is subject. General properties of the equations thus established express the relations that make the laws, to which the theory is applied, depend on one another. Mathematical analysis will demonstrate theorems that state these properties and exactly delimit their import with ultimate rigor.

The consequences of the theory must be submitted to the control of experimentation. In general, theory introduces the consideration of quantities proper to each body, with values that must be determined by measurement. Right up to the last details, mathematical analysis will discuss the particular problems that justify controlled experiments or that serve to institute methods of measurement.

But mathematical analysis may devote itself to demonstrating general theorems, even though these theorems do not serve to establish a connection between experimental laws at all. Or mathematical analysis may exhaust its efforts in resolving particular problems that are useless to the experimenter. In this case, it forgets that mathematical analysis ought to be no more than an instrument in the study of physics. In putting itself forward to the theoretician as an end, it exceeds its role.

It is not that efforts produced in this way are always wasted. In making an instrument better and more complicated than required by the uses to which it will be put, it may happen that the instrument is made ready for other uses. Thus, the theorems that the analyst deduces from certain equations of mathematical physics, though perhaps useless for the theory that furnished the equations, may cast a great deal of light on another theory. Celestial mechanics, for example, led to the study of harmonic functions. Geometers discovered a great many properties that had no applications in celestial mechanics in these functions. But these properties are in continual use in theories of heat, electricity, and magnetism.

Besides, even in the absence of an application, analytical developments in a physical theory may possess a beauty that gives a purpose to mathematics even when it is considered useless. Those who, in perfecting a tool, surpass the requirements of utility to the point of attaining beauty and giving birth to a work of art have certainly not wasted their time and efforts.

But if we must admire those who deduce theorems capable of clarifying another theory from the equations of a physical theory and also those who derive a beautiful analytical system from them, we can only condemn those for whom physics is a pretext to make calculations lacking either utility or beauty. The cleverness of their constructions, the complexity of their combinations, or the subtlety of their intuitions may astonish us for a moment. But then we turn away from their researches with the feeling of regret that all wasted effort inspires. The latter are mechanical artificers who might have been able to construct a practical machine and who have invented only a mechanical curiosity.

10. How Theoretical Physics Is Useful

We have seen the nature of physical theory, what philosophical significance it is suitable to attribute to its results, and the proportion in which experiment and mathematical analysis must be combined in order to construct it. It remains to indicate precisely the kind of utility possessed by the study of this science.

The aim of physical theory is to relate among themselves, or classify, items of knowledge acquired by the experimental method. Without the systematic connection that speculation establishes among them, the laws given by experimentation form a confused and inseparable mass. The human mind needs a thread to guide it in this labyrinth. Theory provides that thread. Theory is therefore devoted to coordinating the laws revealed by experimentation. It is not devoted to revealing new laws.

It sometimes happens that, as a consequence of their deductions, theorists predict an experimental law that had not already been recognized through experimentation. Discoveries of this kind strike the mind vividly, but they are rare. Most experimental discoveries are rightly due to the experimental method. Many physicists criticize a theory for the small number of new facts that it predicts. A more exact knowledge of the proper domain of each kind of research would lead them to admire these predictions. They are the proofs of the fertility of a method that gives more than ought to be required of it.

If theory does not have as its object the discovery of new experimental laws, still less does it have as an object the production of inventions useful in practical life. Theoretical speculations, experimental researches, and practical applications are three distinct domains it is important not to confuse. Those who explore one of these domains are not required to make discoveries in the others. But if these domains are distinct, they are not independent. Knowledge of each of them assists knowledge of the others. A continual exchange of questions and information should be established between the explorers of these different domains.

Practical needs suggest to experimenters phenomena to be observed and laws to be established. The laws established by experimenters provide engineers with principles that they are permitted to modify in order to perfect their inventions. This shows the continual influence of applied science on experimental science and that of experimental science on applied science.

These laws at which experimenters have arrived are the material on which theoreticians work. They classify them. They summarize them in a small number of propositions, which permits the mind to see them as a group and to take hold of their relationships. And when the efforts of theoreticians have thus condensed a great number of laws into a small number of simple symbols that are clear and easy to manipulate, experimenters see clearly in every area of physics what is completed and what remains to be done. Engineers, taking in at a glance the innumerable laws revealed by experimentation, can quickly and surely take hold of those that will be useful to them. Assuredly, those who have made use of the great progress in the electrical industry over the last few years did not create the theory of electricity. But if Pacinotti, Gramme, Siemens, or Edison[9] has been able to manipulate electric current and to place it in the service of human industry, it is because Ampère, Faraday, Ohm, Kirchhoff, Neumann, and Weber have served the human intellect and have shown physicists how to manipulate the laws that current obeys.

Let us remember, therefore, "That it is not useless to attempt to reunite facts under a single point of view by connecting them to a small number of

9. [These figures are connected by their contributions to the design of dynamos. Antonio Pacinotti (1841–1911) and Zénobe Théophile Gramme (1826–1901)

general principles. This is a means of taking hold of the laws more easily, and I think that efforts of this sort can contribute as much as observations themselves to the advancement of science."[10]

---

independently originated the ring method of armature winding. The alternative drum method was invented by Ernst Werner von Siemens (1816–1892) and used in his shuttle-wound armature. Siemens was also the first to realize the advantages of making the poles of the dynamo conform to the armature core's shape as closely as possible. The work of Edison (1847–1931) contributed to the general recognition that Siemen's armatures were superior for practical purposes. His other inventions were so numerous and well known that he became the popular archetype of the nineteenth-century inventor. The list is truly international: Pacinotti was Italian, Gramme Belgian, Siemens German, and Edison, of course, American.]

10. Fresnel, *Oeuvres*, 1, p. 484.

2

Physics and Metaphysics[1]

As Duhem indicates, he is responding to an article by Eugène Vicaire, a Catholic civil engineer, graduate of the Polytechnique and Ecole des Mines and member of the Société mathématique de France. Vicaire wrote "On the Objective Value of Physical Hypotheses," opposing the seeming positivism of Duhem's 1892 article "Some Reflections on the Subject of Physical Theories." Vicaire criticized Duhem's separation of physics and metaphysics for implying the positivistic thesis that physics or positive science was the only real knowledge. He detected in Duhem's views "the poison of skepticism." In his reply, Duhem argues that metaphysics is a real form of knowledge more excellent than physics but separated from it in that it has different objects and is governed by different methods. Duhem's answer fits well into the framework of neo-scholasticism, but it does not go as far as to reunite the disparate forms of knowledge into an overall system of subaltern and subalternated sciences.

Some time ago, we published some reflections on the theories of physics in the pages of this *Review*.[2] We devoted ourselves above all to delineating the exact role of physical theories, which, in our view, are no more than a means of classifying and coordinating experimental laws. They are not metaphysical explanations that reveal to us the causes of phenomena.

---

1. ["Physique et métaphysique," *Revue des questions scientifiques* 34 (1893): 55–83.]

2. P. Duhem, "Quelques réflexions au sujet des théories physique," *Revue de questions scientifiques*, 2nd series, 1, Jan. 1892 [chapter 1 in this volume].

This idea has not been to the taste of all thinkers; several have written in rebuttal of our assertion, rising against it in a lively fashion. Most recently, one of the most justly celebrated members of our scientific Society, Mr. Vicaire, has devoted an article in the *Revue des questions scientifiques* to attacking it.[3]

Without wishing to treat here all the objections raised, explicitly or implicitly, by Vicaire against our perspective, we think that his thesis can be faithfully summarized in the following manner:

> *It is not true that when constructing its theories, positive science has as its object simply to classify experimental laws; its proper object is the discovery of causes. To deny this is to maintain a suspect doctrine of positivism, and one capable of leading to skepticism. That doctrine, condemned by the whole tradition of great physicists, is dangerous, for it destroys scientific activity.*

It is this thesis, opposed to our own, that we propose to attack point by point.

In order to avoid all confusion among those of our readers who are accustomed to the vocabulary of scholastic philosophy, we would like to begin with an important clarification.

To conform to contemporary usage, we give the name *physics* to the experimental study of inanimate things, considered in three phases: the observation of facts, the discovery of laws, and the construction of theories. We regard the investigation of the essence of material things, insofar as they are causes of physical phenomena, as a subdivision of *metaphysics*. This subdivision, together with the study of living matter, forms *cosmology*. This division does not correspond exactly to the peripatetic one. The study of the essence of things constitutes *metaphysics* in peripatetic philosophy. The study of the *motion* of material things—that is, the modifications that the essence of things undergoes in any passage from potential to actual—is *physics*. Peripatetic physics and metaphysics are unified under the name *metaphysics* in our contemporary speech. Peripatetic *physics* is our *cosmology*. Peripatetic philosophy gives no special name to the science of the experimental study of physical laws and their unification in theories. At the time of Aristotle, a single branch of this science, *astronomy*, possessed a development capable of attracting attention. Thus, what we would say in general about *physics*, understood in the con-

3. E. Vicaire, "De la Valeur objective des hypothèses physiques," *Revue de questions scientifiques* 33 (1893): 451–510.

temporary sense, corresponds approximately to what the ancients said about *astronomy*.

I. The Distinction between Physics and Metaphysics

The human intellect does not have direct knowledge or immediate vision of the essence of external things. What we know directly of these things are the phenomena that arise from them and the sequence of these phenomena.

From the knowledge of phenomena, we can draw some knowledge of the things themselves, because they are the efficient causes of these phenomena and because knowledge of an effect provides us with some information on the substance that causes this effect, without giving us, however, a full and adequate knowledge of that substance.

Thus, to acquire an understanding of the external world as complete as our means of knowledge permits, we must ascend successively two degrees of science. We must, in the first place, study phenomena and establish the laws of succession they follow. In the second place, we must induce from these phenomena the properties of the substances that cause them.

The second of these sciences has received the name *metaphysics*. The first is divided into various branches, according to the nature of the phenomena studied. The branch of science which studies phenomena arising from inanimate matter today bears the name of *physics*.

When, in what follows, we speak of metaphysics, we intend always to speak of the part of metaphysics that treats nonliving matter and which, in consequence, corresponds to physics through the nature of the things it studies. That part of metaphysics is often called *cosmology*.

We can summarize what we have just said in the two following definitions:

Physics is the study of phenomena arising from brute matter and of the laws that govern these phenomena.

Cosmology seeks to understand the nature of brute matter, considered as the cause of phenomena and as the foundation (raison d'être) *of physical laws.*

Hence, there exists a distinction in kind between metaphysics and physics.

It is important, however, not to be mistaken about the origin of this distinction. It does not follow from the nature of the things studied, but only from the nature of our intellects. An intellect which had a direct intuitive view of the essence of things—such as, according to the teaching of the theologians, an angel's intellect—would not make any distinction between physics and metaphysics. Such an intellect would not know successively the phenomena and the substance—that is, the cause of these phenomena. It would know substance and its modifications simultaneously. It would be much the same for an intellect that had no direct intuition of the essence of things but an adequate—though indirect—view through the beatific vision of divine thought.

II. That Physics Logically Precedes Metaphysics

The knowledge that metaphysics gives us of things is more intimate and deeper than the one provided by physics. It therefore surpasses the latter in excellence. But if metaphysics precedes physics in order of excellence, it comes after physics in the order of logic. We cannot come to know the essence of things except insofar as that essence is the cause and foundation for phenomena and the laws that govern them. The study of phenomena and laws must therefore precede the investigation of causes. In the same way, when one ascends a staircase, the highest step is the one crossed last.

In order to avoid any misunderstanding, we must insist on this logical priority of physics over metaphysics as an essential point.

Here, to begin with, is a proposition that it seems to us cannot be contested: Any metaphysical investigation concerning brute matter cannot be made logically before one has acquired some understanding of physics.

It is quite evident, in fact, that one cannot think of investigating anything whatsoever about the causes of phenomena without having studied the phenomena themselves and having acquired some understanding of them.

But once some knowledge of physics has permitted the first metaphysical investigations and these investigations have provided some indications about the nature of material things, can one not follow the inverse order, descending the staircase one has climbed, and, from what one knows about the nature of material things, deduce the phenomena which they must produce and the laws that these phenomena obey?

To deny in an absolute manner the possibility of such an intellectual path seems to us rash at the minimum. Theoretically, it is possible that the knowledge of the nature of things, obtained through metaphysics, permits the establishment, by deduction, of a true physics. But practically, the method that consists of taking metaphysics as the point of departure in the discovery of physical truths appears very difficult and full of danger. It is easy to reveal the reason for this.

A complete and adequate knowledge of substances carries with it a complete and adequate knowledge of the phenomena they can produce. The knowledge of causes implies the knowledge of effects. But the reverse of this proposition is not true. The same effect can be produced by several different causes. To this extent, even the total and complete knowledge of a set of phenomena would not give us a complete knowledge of the substances through which they are produced.

Thus, when we ascend from effects to causes in order to obtain a metaphysics, starting from some established physical knowledge, as perfect and extensive as one would like, we gain a very incomplete and imperfect knowledge of the essence of material things. This knowledge proceeds more through negation than through affirmation, more by the exclusion of some hypotheses that might be made about the nature of things than by positive indications of that nature. It is only in certain rare cases, through the exclusion of all possible hypotheses except one, that we are able to acquire positive proof about the essence of material things.

To understand this essential point properly, it is important never to confuse the *truths established by metaphysics* with *metaphysical systems.* The truths of metaphysics are propositions few in number and, for the most part, negative in form, which we obtain in ascending from observed phenomena to the substances which cause them. A metaphysical system, however, is a collection of positive judgments—although hypothetical for the most part—by means of which a philosopher seeks to relate metaphysical truths among themselves in a logical and harmonious order. Such a system is acceptable provided none of the hypotheses composing it conflicts with an established metaphysical truth. But it remains always highly problematic and never forces itself on reason in an unavoidable fashion.

What we have just said on the subject of metaphysical truths makes evident that these truths can almost never become the point of departure for a deduction leading to a physical discovery. When, by depending on knowledge of a set of phenomena, we have succeeded in demonstrating the impossibility of certain assumptions concerning the substances through which the phenomena are produced, in acquiring even some positive indications on the subject of these substances, the view we have of them remains too general and too little *determinate* to enable us to foresee the existence of a new class of phenomena or to anticipate a new physical law.

Metaphysical systems present to us a definition of the nature of things more detailed and more determinate than that furnished by demonstrated metaphysical truths. Because of that, metaphysical systems become capable of leading us to physical consequences more easily than can metaphysical truths alone. But while a physical consequence deduced from some metaphysical propositions participates in the certainty of the latter, a physical consequence deduced from a metaphysical system suffers from the doubtful and problematic character affecting the system itself and cannot be regarded as established. It is no more than an indication that physics will have to examine and on which physics will rule.

In conclusion, *if it is not impossible, it is at least extremely difficult to deduce a new physical truth from well-established metaphysical truths. As for metaphysical systems, they may suggest a proposition in physics, but physics alone can decide if this proposition is correct or incorrect.*

III. Physics Rests on Principles Evident in Themselves and Independent of Any Metaphysical Considerations

Since it is impossible—if not in theory, then at least in practice—to call forth any new physical truth from metaphysical knowledge that we are able to acquire about the nature of things, physics must necessarily be able to constitute itself through a proper method independent of any metaphysics. This method, which permits the study of physical phenomena and the discovery of the laws that connect them, without recourse to metaphysics, is the *experimental method.*

This method employs a certain number of concepts, for example, the concepts of physical phenomenon and physical law, body, extension, time, and motion. It rests on certain principles, such as the axioms of geometry and kinematics and the existence of laws determining the connection of physical phenomena.

To use these concepts, to make use of these principles, it is not necessary to have constructed a metaphysics. These principles and concepts appear to our intellect sufficiently certain and sufficiently distinct in themselves that we should be able, without fear of confusion or error, to put them into play through the experimental method. In fact, a good number of physicists juggle these concepts and principles with sureness, precision, and fruitfulness, as foundations of the science that they deepen and develop, without asking themselves for a single instant what a body is or what a law is from the metaphysical point of view.

It is in this sense that one may state the following proposition: *The experimental method rests on principles evident in themselves and independent of any metaphysics.*

It does not follow from this that the foundations of the experimental method escape the grasp of metaphysics and cannot become objects of study for that science. Apart from any metaphysical investigation, we have the concept of body and the concept of law in a manner distinct enough to be able to make a legitimate use of these concepts in all physical investigations. Apart from any metaphysical investigation, we know that the phenomena arising from matter are subject to fixed laws, and this principle is so certain that we are able, without hesitation, to dedicate our life to the discovery of these laws. But from the fact that we have knowledge of these concepts and sufficient confidence in this principle to enable us to make use of these concepts and this principle in the course of our experimental investigations, it does not follow that this knowledge is absolutely clear and complete, or that the foundations on which it rests are known to us, or that we have nothing to learn about these questions. For example, we have a sufficient idea of body that we would be confident in not taking for a body something that was not one. It does not follow from this that we would know in a complete and adequate way what a body consists of. It does not follow from this that we would be forbidden to investigate, within the bounds of possibility, and to weigh through metaphysical examination the foundations of the experimental method in order to penetrate its essence and foundation.

But this metaphysical investigation, however important it might be in its own right, has no repercussions for the experimental method. In seeking to give ourselves a metaphysical account of one of these concepts or of one of these principles on which physics rests, we would not modify in the least the use which is made of this concept or principle in physics. Place side by side a philosophical physicist who has used his evenings to delve into the metaphysical concept of body and another physicist who is devoted exclusively to his science and who has never reflected for five minutes on this same concept. Both of them make the same use of this concept in experimental practice. What is self-evident in this concept is what is necessary and sufficient in physics. What metaphysics discovers about it afterwards is absolutely useless to those who are, and wish to be, no more than physicists.

Thus, *metaphysics aims to give an account of the self-evident foundations on which physics rests. But this study adds nothing to their certainty and to their clarity in the domain of physics.*

IV. Physical Theories Are Independent of Metaphysics and *Vice Versa*

All experimental science is composed of at least two phases: the observation of facts and their reduction to laws. But in those that, like physics, have arrived at a sufficient degree of development, a third phase is conjoined to the two others. This is the theoretical phase. Without it, experimental laws would form a confused mass impossible to disentangle, where the mind would have extreme difficulty in finding its bearings and in discovering the law it needed to use in each particular case. *The aim of theory is to classify experimental laws.* Between a set of experimental laws taken as experimentation has brought them to light and the same set of laws connected by a theory, there is the same difference as that between a mass of documents heaped in confusion and the same documents carefully classified in a methodical collection. They are the same documents; they say exactly the same thing and in the same way. But in the first case, their disorder makes them useless, for one is never sure of recovering the document one needs at the moment one needs it; similarly, in the second case, the documents are made fruitful by a methodical grouping which places the desired document surely and without effort in the hands of the researcher.

Physical laws retain exactly the same sense when a theory connects them as when they are dispersed and isolated. They teach us no more in the first case than in the second. In the first case, however, they are easier to encompass and easier to use than in the second. Thus, physical science does not change its character or its content in becoming theoretical. It becomes more perfect in form, better ordered, simpler, and, in consequence, more beautiful. It remains the same at its foundation. It remains physics; it does not become metaphysics. *In classifying a group of experimental laws, physical theory teaches us absolutely nothing about the foundation for these laws and the nature of the phenomena that they govern.*

Thus understood, and thus reduced to its true role, physical theory becomes like the whole of physics, absolutely independent of metaphysics. Since none of the propositions, which taken together constitute a physical theory, is a judgment about the nature of things, none of these propositions can ever be in contradiction with a metaphysical truth, which itself is always a judgment on the nature of things. This essential difference between a proposition of theoretical physics and a metaphysical truth shows equally that the one can never be identical to the other. *It is therefore absurd to seek among the truth of metaphysics either the confirmation or the refutation of a physical theory, at least to the extent that it remains confined to its proper domain.*

Reciprocally, since in classifying a set of laws a physical theory adds absolutely nothing to the content of those laws, it provides, as points of departure for metaphysical investigation, no data other than those that one could draw from the same laws before classification or reduction to theory. In consequence, insofar as physical laws are the logical point of departure for all metaphysical research on the essence of material things, physical theories would not be able to exert any direct influence on the progress of this research. If such theories serve metaphysics, it is indirectly—in making the laws they classify and condense more easily presented to the mind of the philosopher. *The subordination that a theory establishes among various physical laws by classifying them does not oblige us to admit a similar subordination among the metaphysical laws of which the physical laws are the manifestation.*

One might summarize the two propositions that we have just established by saying that *physical theories and metaphysical truths are independent of one another.*

Since this is the essential point of our discussion, let us give some more clarifications, in order to avoid any misunderstanding.

Let us imagine that we have come to a profound, detailed metaphysical knowledge of the essence of material things. The physical laws that follow from this essence would appear to us in an order or subordination which results from their nature itself. It is certain that this order would give us the best classification of these physical laws. It is certain that a complete metaphysical explanation of the nature of material things would provide us, by the same token, with the best of physical theories. But let us be careful to note, even when we have knowledge of this physical theory, a reproduction of the metaphysical order, that we would still be free logically to adopt another, to connect physical laws in a different order, to accept another mode of representation of physical phenomena. No doubt it would be unreasonable to reject the first theory, because it is the best. We would transgress the law telling us that in every order of things we should choose what excels. But we would not violate any principle of logic. We would not commit an absurdity. A classification, in fact, is not a judgment. It can be convenient or inconvenient, good or bad; it cannot be true or false.

Besides, the hypothesis under which we have just placed ourselves is purely ideal. Our certain metaphysical knowledge, as we have seen, is too indeterminate, too negative in character, to indicate for us in what order the various physical laws are subordinated one to another, or to give us from these laws a classification capable of being erected on physical theory. In order to deduce a definite physical theory from metaphysical principles, one must depend not only on demonstrated metaphysical truths but also on a metaphysical system. And in fact, there is almost no metaphysical system that has not sought to establish physical theories. But a metaphysical system, however acceptable, however satisfactory one supposes it to be, is always hypothetical to a high degree. It is therefore in no way evident that a physical theory deduced from a metaphysical system would be better than another theory established apart from all consideration about the essence of things.

So, to the extent that it remains in its proper domain and confines itself only to classifying experimental laws, a physical theory is absolutely independent of all metaphysics. And not only does it not depend on the more or less probable metaphysical systems disputed by the philosophical schools, but it is also independent of the better established metaphysical truths concerning the essence of material things, to the extent that it remains equally acceptable not only for those who support the most varied metaphysical systems but also for those who deny the best-demonstrated metaphysical truths. Encamped in its fort, it need fear only two kinds of opponents: physicists who dispute it in the name of experiment or in the name of other physical theories, and skeptics who deny the evidence and certainty of some of the concepts and some of the principles on which experimental science logically depends. Physics has no power to fight the latter; it is not equipped for that. It is for metaphysics to show that the foundations of the experimental method are firm. The physicist is constrained to admit this truth as evident. In the proper field of his theories, the physicist cannot and must not accept conflict except with a physicist.

V. The Thesis Presented Above Is Neither Skeptical Nor Positivistic

We have just presented the essential thesis, in our view, concerning the independence of physical theories and metaphysical investigations. Let us try now to dispel some of the objections most often addressed to this thesis.

Does affirming the natural separation that exists between physical theories and metaphysical doctrines open a door for skepticism? Does it make a concession to positivism?

It is almost impossible to delimit the correct boundaries of a science, those imposed on it both by the nature of the objects it studies and by the nature of our minds, without immediately being accused of skepticism. It seems to some people that each of the logical methods our reason deploys is all powerful, that each of them can engage all subjects and show in them the most hidden secrets. In the workshop of human understanding, each tool is appropriate, according to them, for the most varied tasks, and our intellect resembles a little that chemist who boasted of knowing how to file with a saw and saw with a file. What a deadly pretension of dogmatism, which engenders the worst errors and furnishes skepticism with its most troubling arguments! Question a spirit gnawed by doubt—not that facile and shallow doubt born from laziness and vanity, but the anxious and painful doubt that comes from analysis and meditation. Seek the path by which the doubt entered this spirit. Ask it how its faith in reason has vanished. Always you will receive a similar response. Always [the spirit] has despaired because deductions carefully laid out led it to a manifestly false conclusion, or because a demanding investigation has refused to produce an expected result. Consider then the source of this error or of this sterility. Always [it is] an illegitimate extension given to a legitimate method of logic. The tool has been prepared for a definite kind of work; the tool user wished to give it another. Manipulating it for a long time, using force, bringing his dexterity to bear, has had no result or no result except drudgery. So, rebuffed, he has thrown the tool away and folded his arms.

Would you wish to lead these discouraged people back to their labors? Or would you rather, from the outset, enable them to avoid miscalculations and disappointments? Show them the proper use of their tools. Show them that the saw is no good except for sawing and the file no good except for filing. Matters are the same with the means of understanding that God has placed at the disposal of our reason. Nothing is more favorable to skepticism than to mix up the domains of various sciences. By contrast, nothing is more efficient against this tendency toward blurring than the exact definition of the diverse methods and the precise demarcation of the field that each one of them must explore.

In denying metaphysics the right to govern physical investigations, and in denying to physical theories the right to erect themselves into metaphysical explanations, are we being positivistic? We hold that the positive sciences must be treated by methods proper to positive sciences. We hold that these methods rest on principles that are evident in themselves and able to function independently of any metaphysical investigation. We hold that these methods, which are effective in the observation of phenomena and in the discovery of laws, are incapable of capturing causes and reaching substances. But this is not to be a positivist. To be a positivist is to assert that there is no logical method other than the method of the positive sciences, that anything that cannot be achieved by this method, anything unknowable to positive sciences, is in itself absolutely unknowable. Is this what we are maintaining?

Do you wish to play the game of positivism? Confuse the domain of metaphysics with the domain of physics, the metaphysical method with the experimental method. Discuss physical theories using reasons derived from metaphysical systems. Engulf the theories of positive science in your metaphysical systems. The positivist would have no difficulty in demon-

strating to you that physical methods are not able to obtain the consequences that you aspire to deduce from them, and he would conclude from
this failure that the foundations of metaphysics totter. And the positivist
would have no difficulty demonstrating to you that your metaphysical
deductions are able to do nothing in the field of physical theories depending upon experimental laws and would conclude from it that metaphysics
is condemned through its consequences.

If you do not establish a radical separation between physics and metaphysics, if you mix them together, you are bound to recognize that physical method is good even in metaphysics. This is to give comfort to the
cause of positivism.

VI. The Preceding Thesis from the Viewpoint of Tradition

Those who argue against the preceding thesis willingly claim that they
depend on tradition. As they put it, all the great thinkers and all the great
scientists have considered physical theories as an attempt, or as a step
toward, the metaphysical explanation of things. All have sought not to
classify phenomena but to discover their causes. It is the hope of disclosing
the reasons for physical effects which has given them the courage to pursue
their investigations, and the fertility of the latter gives us evidence that this
hope was not an illusion.

From the historical point of view, nothing is more false than this manner of envisaging tradition.

On the subject of the relations between physics and metaphysics, Aristotle and the peripatetic philosophy admitted a thesis which essentially
agrees with the one we have developed. They made little use of it except
in astronomy, the only branch of physics which was developed at that
period, but what they said about the motion of the stars can be extended
readily to other natural phenomena.

> They clearly separated *astronomy*, the science of celestial *phenomena*, from
> investigations of the *causes* of the motions of the stars and speculations on the
> reality or unreality of these motions. Studies of this type were reserved for
> *physics*, that is to say that part of philosophy today called *cosmology*. From then
> on, choosing between astronomical hypotheses was for them a matter of indif-
> ference, and there was nothing inconvenient about adopting the geocentric
> viewpoint, which conformed to the appearances better and was easier to apply
> than the alternative.[4]

4. P. Mansion, *Sur les principes fondamentaux de la géometrie, de la méchanique et
de l'astronomie* (Paris, 1893).

On this subject, Schiaparelli cites a characteristic passage from Posidonius (or his abbreviator, Geminus), preserved by Simplicius, Aristotle's commentator:

> It is unimportant for the astronomer to know what is immovable and what is moved. He is able to admit every hypothesis that represents the phenomena, for example, the one proposed by Heraclides Ponticus, according to which the mean anomaly of the planets in relation to the sun is explained by a motion of the earth around the sun, considered as fixed. The astronomer must then have recourse to the physicists for the fundamental principles of his investigations.

All Aristotle commentators adopt the opinion so clearly expressed in the passage that we have just cited. Thus, St. Thomas Aquinas, in his commentary on Aristotle's *De Caelo*, expresses himself in the same way on the hypotheses of astronomers:

> The assumptions of those people (the astrologers) are not necessarily true. Although they save the appearances by suppositions constructed in this way, one ought not to say that these suppositions are true, because one might save the appearances concerning the stars equally well by means of some other method not yet understood by men. However Aristotle speaks this way about the truth of suppositions on the quality of motion.[5]

It is not only the philosophers of antiquity and the Middle Ages who separate purely representational hypotheses without metaphysical import, which physicists use to classify facts, from the true explanation of the same facts. Astronomers and physicists conform to these principles in their writings.

For example, when Archimedes undertakes to write a mathematical theory of floating bodies—the first theory of mathematical physics ever composed—he does not seek to understand what liquids are in themselves and to uncover the metaphysical foundation of their properties. He contents himself with stating a proposition which he names a *hypothesis*, and with demonstrating that the physical laws of floating bodies can be deduced logically from that hypothesis. That fundamental hypothesis of Archimedes can be stated in the following manner:

5. "Illorum (Astrologorum) autem suppositiones quas adiuvenerunt, non est necessarium esse veras: licet enim talibus suppositionibus factis appareant solvere. non tamen opportet dicere has suppositiones esse veras, quia forte secundum aliquem alium modum nondum ab hominibus comprehensum apparentia circa stellas salvatur. Aristoteles tamen utitur huiusmodi suppositionibus ad qualitatem motuum tanquam veris." *Ad. Lect.* XVII, book 2, ii.

Let us suppose a liquid is of such a nature that, its parts lying evenly and being continuous, that part which is thrust less is driven along by that part which is thrust more; and each of its parts is thrust by the liquid above it in a perpendicular direction if the liquid is sunk in anything and compressed by anything else.[6]

One sees clearly, by the very nature of this hypothesis, that it does not aspire to be a metaphysical explanation of the properties of liquids. The foundation for these properties does not become more apparent in any fashion when Archimedes shows that one can draw them all out logically from the preceding proposition. Only these properties are thus classified and condensed. To this extent, the first theory ever written in mathematical physics is at the same time a model for theories in the sense we intend.

Copernicus proceeds in astronomy as Archimedes did in hydrostatics:[7]

In our time we have recovered a sort of summary or announcement for his book the *Revolutions*, a summary he wrote around 1530. The title of this short work is: *A little book of Nicolas Copernicus on the hypotheses of the celestial motions put together by him*.[8] In the preamble, he announces that he is going to explain the system of the world better than his predecessors: "If our different assumptions, called axioms, are admitted."[9] There follow seven postulates, where he asks that we concede to him that the sun is stationary, that the earth moves, that the stars are at enormous distances, etc.

In the *Narratio Prima* of Rheticus,[10] a more extended announcement written under the inspiration and no doubt before the eyes of Copernicus, there is no question throughout except of hypotheses ancient or new.

6. "Supponantur humidem habens talem naturam ut partibus ipsius ex aequo jacentibus et existentibus continuis expellantur minus pulsa a magis pulsa, et unaquaeque autem partium ipsius pellatur humido quod supra ipsius existente secundum perpendicularem."

7. The following, concerning Copernicus, follows P. Mansion, loc. cit.

8. *Nicolai Copernici de* HYPOTHESIBUS *motuum coelestium a se constitutis commentariolus.* [This book is now typically known by the abbreviated title *Commentariolus.* See Noel M. Swerdlow, "The Derivation and First Draft of Copernicus' Planetary Theory: A Translation of the *Commentariolus* with Commentary," *Proceedings of the American Philosophical Society* 117 (1973): 423–512.]

9. Si nobis aliquae PETITIONES, quas axiomata vocant, concedantur.

10. *Georgii Joachimi Rhetici Narratio Prima*, Danzig, 1540. [The best modern version is Hugonnard-Roche, H., and Verdet, J.-P., eds. and trans. (1982) *Georgii Joachimi Rhetici Narratio Prima. Studia Copernicana*, XX. Warsaw: Ossolineum, a critical edition and French translation, with copious notes.]

Things are the same in the book *On the Revolutions*. Copernicus leaves the ground of astronomy to engage that of physics in the Aristotelian sense, that is to say cosmology, in only two chapters. In one (book 1, chapter 7), he presents Ptolemy's reasons in favor of the immobility of the earth. In the other (chapter 8), he tries to show that they are not compelling when taken in the context of physics. He concludes modestly: "You see therefore that from all this it is more probable that the earth moves than that it stays still. . . ."[11]

But in the whole of the rest of the work, he writes from the phenomenal point of view. He is satisfied with giving systematic explanations of the celestial motions, the immobility of the sun being conceded, or on the assumption that the earth moves, as he puts it on many occasions.

The author of the anonymous preface to *On the Revolutions*, probably Osiander, has therefore at the same time summarized Thomist tradition and the thought of Copernicus, rather than betraying it, as has often been said, in writing the following passages: "It is not necessary that these hypotheses be true, they need not even be likely. This one thing suffices: that the calculation to which they lead agrees with the result of observations. . . . Let no one then expect from astronomy any doctrine about these hypotheses that is certain. Astronomy can give nothing of the sort."[12]

At the end of the sixteenth century and the beginning of the seventeenth, the human mind underwent one of the greatest revolutions ever to turn the world of thought upside down. The logical rules, delineated by the genius of Greece, had been accepted until then with intelligent docility by the masters of the Schools, and then with strict servility by scholasticism during its decadence. At this moment, thinkers rejected them. They aspired to reform logic, to forge anew the tools which human reason uses, and, with Bacon, to create a *novum organum*. They shattered the lines of demarcation established by the peripatetics between the different branches of human knowledge. *Distinguo*, which served to delimit questions exactly and indicate to each method the field proper to it, became a term of ridicule used in comedy. Then one saw the disappearance of the ancient barrier separating the study of physical phenomena and their laws from the investigation of causes. Then one saw physical theories taken for metaphysical explanations and metaphysical systems seeking to establish physical theories by deductive means.

11. Vides ergo quod ex his omnibus, PROBABILIOR sit mobilitas terrae quam euis quies, praesertim in quotidiana revolutione, tanquam terrae maxime propria.

12. Neque enim necesse est eas hypotheses esse veras, imo ne versimiles quidem, sed sufficit hoc unum si calculum observationibus congruentem exhibeant. . . . Neque quisquam, quod as hypotheses attinet, quicquam CERTI ab astronomia expectet, cum ipsa nihil praestare queat [emphasis added].

The illusion that physical theories attain true causes and ultimate reasons for things penetrates in every sense the writings of Kepler and Galileo. The debates that make up the trial of Galileo would be incomprehensible to anyone who did not see there the struggle between a physicist who wishes his theories to be not only the representation but also the *explanation* of phenomena, and theologians who maintain the ancient distinction and do not admit that Galileo's physical and mechanical reasonings might in any way go against their cosmology.

But the person who made the greatest contribution to breaching the barrier between physics and metaphysics is Descartes.

Descartes's method calls into doubt the principles of all our knowledge and leaves them suspended by this methodical doubt until the moment their legitimacy can be demonstrated by a long chain of deductions beginning with the celebrated "I think, therefore I am." Nothing could be more contrary to the peripatetic idea than such a method, according to which a science such as physics rests on self-evident principles whose nature metaphysics can unearth but whose certainty it cannot establish.

The first proposition in physics that Descartes establishes, by following his method, gives him, he tells us, knowledge of the very essence of matter. "The nature of body consists in this alone—that it is a substance that has extension." The essence of matter being thus known, one should be able, through the method of geometry, to deduce from it the explanation of all natural phenomena. "I accept no principles in physics," says Descartes, summarizing the method through which he attempts to treat that science, "that would not also be accepted in mathematics, in order to be able to prove by demonstration everything that I would deduce from them, and these principles are sufficient, to the extent that all the phenomena of nature can be explained by their means."

Such is the audacious formula of Cartesian cosmology. Human beings know the very essence of matter, which is extension. They are therefore able, logically, to deduce from it all the properties of matter. The distinction between physics, which studies phenomena and their laws, and metaphysics, which seeks to grasp some information about the essence of matter insofar as it is the cause of phenomena and the foundation of the laws, no longer has any basis. The mind does not begin from knowledge of phenomena to raise itself in turn to knowledge of matter. What the mind knows first of all is the very nature of matter and the explanation of phenomena that flows from it.

Descartes pushes the consequences of this bold formula to the limit. He is not satisfied with asserting that the explanation of all natural phenomena may be deduced from the proposition that "the essence of matter consists in extension." He attempts to give this explanation in detail. He seeks to construct the world starting from this definition, and when his work is complete, he stops to contemplate it and states that nothing is lacking in it. As the title of one of the paragraphs in the *Principles of Philosophy* tells us: "There is no phenomenon of nature which has been overlooked in this treatise."[13]

Descartes, however, seems to have been frightened for a moment by the boldness of his cosmological doctrine and to have sought to reconcile it with peripatetic doctrine. This follows from reading one of the articles of the book *Principles of Philosophy*. Let us quote in its entirety this article, which is so relevant to the object that occupies us:

> However, although this method may enable us to understand how all the things in nature could have arisen, it should not therefore be inferred that they were in fact made in this way. Just as the same craftsman could make two clocks which tell the time equally well and look completely alike from the outside but have completely different assemblies of wheels inside, so the supreme craftsman of the real world could have produced all that we see in several different ways. I am very happy to admit this; and I shall think I have achieved enough provided only that what I have written is such as to correspond accurately with all the phenomena of nature. This will indeed be sufficient for application in ordinary life, since medicine and mechanics, and all the other arts which can be fully developed with the help of physics, are directed only toward items that can be perceived by the senses and are therefore to be counted among the phenomena of nature. And in case anyone happens to be convinced that Aristotle achieved—or wanted to achieve—any more than this, he himself expressly asserts in the first book of the Meteorologica, at the beginning of Chapter Seven, that when dealing with things not manifest to the senses, he reckons he has provided adequate reasons and demonstrations if he can simply show that such things are capable of occurring in accordance with his explanations.[14]

But this sort of concession to the ideas of the schools is manifestly not in accord with Descartes's method. It is only one of the precautions that the great philosopher took willingly against the censure of the Church, strongly motivated, as one knows, by the condemnation of Galileo. For the rest, it seems that Descartes himself might have feared that one might take his caution too seriously, for he follows the article we have just cited with

13. [*Principles* IV, art. 199.]
14. [*Principles* IV, 204.]

two others, titled as follows: "Nevertheless my explanations appear to be at least morally certain"; "Indeed, my explanations possess more than moral certainty."[15]

We therefore think that one cannot look at Descartes correctly except as the first philosopher who stopped discriminating physics from cosmology, or at least as the one whose writings most clearly and completely denied the distinction between these two orders of knowledge.

The influence of Descartes on the great minds of his century was immense. Also we see, following him, that the most powerful physicists regard their theories as true explanations of the nature of things and apply them through reasons derived from metaphysics. We have cited elsewhere[16] various passages from Christiaan Huygens that show clearly to what degree he shared Descartes's ideas in this regard.

This influence of Descartes was extremely general. It was not, however, entirely universal. We have shown, in the article to which we have already alluded, that Pascal did not submit to it without some protest. We have shown above all that Newton never abandoned the tradition of the schools. He always clearly separated scientific theories intended to coordinate physical laws and metaphysical investigations intended to make known the causes of phenomena. He always maintained the logical priority of the first, among which he placed celestial mechanics, over the second. By a happy coincidence, in the same journal, Mr. de Kirwan,[17] in commenting on Newton's idea about action at a distance, came to understand the thought of the author of *Mathematical Principles of Natural Philosophy* in the same manner that we did.

In the eighteenth and nineteenth centuries, the exact concept of the relations between physics and metaphysics was more and more obscured. Many reasons, among which the more or less direct influence of Descartes's ideas plays a preponderant role, tended to confuse theories with explanations. One did not need to believe, however, that all trace of the distinction that needs to be made between these two kinds of knowledge had disappeared from the mind of physicists. The same people whose pride of discovery carried the furthest, those who had the most complete confidence in the power of physical theories, recognized, when their med-

---

15. [*Principles* IV, arts. 205 and 206.]

16. P. Duhem, "Une nouvelle théorie du monde organique," *Revue des questions scientifiques,* 2nd series, 3., Jan. 1893, p. 117.

17. C. de Kirwan, "Newton et l'action à distance," *Revue des questions scientifiques,* 2nd series, 3., Jan. 1893, p. 169.

itations paused on this question, that the theories of which they were so proud were perhaps no more than metaphysical explanations.

In an article to which we have just referred the reader, we have cited Laplace among those who regarded the theory of universal attraction as the ultimate explanation of natural phenomena. And in fact, if one excludes the writings of Poisson, there is perhaps no work that exudes a more complete confidence in the power of mathematical theories than the *Méchanique céleste* (*Celestial Mechanics*). This confidence, however, is not entirely blind. In several places in his *Exposition du système du monde* (*System of the World*), Laplace indicates that this universal attraction which in the form of gravity or molecular attraction connects all natural phenomena, is perhaps not the explanation of them—that universal attraction, itself depends perhaps on a higher cause. This cause, it is true, Laplace seems to cast into a domain that is unknowable, but in any case he recognizes no less than Newton that research on this cause, if it is possible, constitutes a problem distinct from that resolved by astronomical theories. "This principle, he says, "is it a primitive law of nature? Is it not a general effect of an unknown cause? Here the present ignorance of the inner properties of matter brings us to a halt, and we remove all hope of responding in a satisfactory manner to these questions."[18] "The principle of universal gravitation," he says again, "is it a primitive law of nature? Or is it not the general effect of an unknown cause? Can one not lead back the affinities to this principle? Newton, more circumspect than many of his followers, said nothing about these questions, about which the present ignorance of the inner properties of matter prevents us from responding in a satisfactory manner."[19]

We have equally cited Ampère among the number of those who thought to find the true explanations of physical phenomena in attractions and repulsions of various natures. It is certain that Ampère regards the laws established by Newton, by Coulomb, and by himself as furnishing at the same time physical theories and metaphysical explanations. But if he believes that he possesses the simultaneous solution of the physical problem and the metaphysical problem, he does not confuse these problems. He encourages those who refuse to recognize the legitimacy of the solutions that he proposes in the domain of metaphysics to accept them at least in the domain of physics:

18. Laplace, *Exposition du système du monde,* book IV, chap. XVII.
19. Laplace, *Exposition du système du monde,* book V, chap. V.

The principal advantage of the formulas which are thus derived immediately from various general facts, given through a sufficient number of observations the certainty of which cannot be disputed, is that they remain independent as much of the hypotheses that aided their authors in the search for these formulas as [also] from those which may later be substituted for them. The expression for universal attraction deduced from Kepler's laws does not depend at all on the hypotheses that various authors have tried to make on a mechanical cause that they wished to assign to it. The theory of heat rests in reality on general facts given immediately by observation, and the equation deduced from these facts is confirmed by the agreement of the results one derives from it and those given by experiment. [This equation] must be equally accepted as expressing the true laws of the propagation of heat, both by those who attribute it to a spreading out of caloric particles and by those who to explain the same phenomena have recourse to vibrations in a fluid spread out through space. It is necessary only that the former show how the equation that concerns them results from their manner of seeing things and that the latter deduce it from general formulas of wave motion. This is not to say anything about the certainty of this equation, except insofar as their respective hypotheses can subsist. The physicist who accepts neither of these positions admits this equation as an exact representation of the facts without worrying about the manner in which it follows from one or the other of the explanations of which we have spoken.[20]

We could multiply these quotations, but those we have given suffice to clarify the idea that we wish to illuminate. Newton, Laplace, and Ampère have shown us that even in modern times, which are proud of the developments of positive science, the healthy and prudent tradition of the schools has never disappeared completely, that those physicists who were greatest because of their discoveries have always recognized that mathematical theories have as their object to coordinate natural laws, and that the discovery of causes constitutes a separate problem, logically posterior to the former. In consequence, [they have recognized that] this doctrine, rather than being pernicious for scientific research, is imposed effortlessly on minds that are the most fertile in discovery.

Should it be said that [this doctrine] has never been misunderstood by the great scientists? Assuredly not. The examples of Descartes and Huygens show us that one may give a prodigious impulse to physical theories while being quite wrong about their nature and confusing them with cosmological explanations, that one can even draw a powerful and fertile

---

20. André-Marie Ampère, *Théorie mathématique des phénomènes électrodynamiques, uniquement déduite de l'expérience*, Paris: Hermann, p. 3.

ardor for scientific research from this error, which exaggerates the importance of the aim to be attained. But these examples hold nothing that can surprise us and that might be capable of breaking down the distinction that we have tried to establish between the construction of a physical theory and the investigation of metaphysical inquiry into causes. Often illusion inflames human activities more than the clear understanding of the object pursued. Is this a reason for confusing illusion with truth? Admirable geographical discoveries have been made by adventurers seeking the Land of Gold. Does this mean that our maps should include El Dorado?

3

The English School and Physical Theories: On a Recent Book by W. Thomson[1]

This article and the next one, entitled "Some Reflections on the Subject of Experimental Physics," form the core of Duhem's most famous work, The Aim and Structure of Physical Theory, *which Duhem published as a series of articles in 1904 and 1905. The themes of this article are reworked in part I, chapter 4, of that larger work.*

A collection of scientific lectures given by William Thomson in various circumstances and bearing on diverse questions of general physics has just been translated into French.[2] In running through these lectures one experiences a very strange feeling—the feeling that one really has before one's eyes the work of a first-rank scientist and that, nevertheless, this work is not altogether science, or at least science as we understand it and as we like it.

We have experienced this feeling to a more or less intense degree every time we have opened a book written by one of the physicists of the current English school: Maxwell or Lodge, Tait or Thomson.[3] It is the special

---

1. ["L'Ecole anglaise et les théories physiques," *Revue des questions scientifiques* 34 (1893): 345–378. Duhem uses "English" (*Anglais*) in the old generic sense to cover all inhabitants of the British Isles, although many of the physicists discussed in this article are Scottish or Irish by birth. See the biographical sketches in subsequent notes.]

2. Sir W. Thomson (Lord Kelvin), "Conférences scientifiques et allocutions," trans. L. Lugol, with notes by M. Brillouin, in *Constitution de la matière* (Paris, 1893).

3. [James Clerk Maxwell (1831–1879) became the first professor of experimental physics at Cambridge (1871–1879), although he was born, and spent a great part

form in which British genius conceives and realizes physical science that causes this feeling of astonishment in French intellects.

It has seemed interesting to us to analyze the causes of this astonishment, to investigate the characteristics of the English scientific intellect, and to classify the features that distinguish "this great English school of mathematical physics, the works of which are one of the glories of the century."[4]

No one better personifies this school than W. Thomson. As ingenious as Faraday, as bold as Maxwell, he is more complete than either of these two geniuses. As skillful an experimenter as the former, he handles geometry as easily as the latter, and surpasses him in spirit of invention in that branch of science. He is not content to encompass the entire field of physical theory, and his researches shine in the realm of practical applications. Thanks to him, navigators are protected from compass errors and undersea cables carry the thought of one continent to another.[5] In addition, the *Scientific Lectures* of W. Thomson supply us with valuable evidence. Through them we will take hold of the scientific intellect of the English in its highest and most perfect form.

---

of his life, in Scotland. His most important work was the *Treatise on Electricity and Magnetism* (1872).

Oliver Lodge (1851–1940) was the first professor of physics at the University College in Liverpool, England, from 1881 to 1900, and thereafter, until his retirement in 1919, first principal of the University of Birmingham. He was knighted in 1902.

Charles Guthrie Tait (1831–1901) became professor of mathematics at Queens College, Belfast, Ireland, in 1854, and professor of natural philosophy at the University of Edinburgh from 1860 until his death. Tait collaborated with Thomson on a multivolume survey of physics, *A Treatise on Natural Philosophy*, which began to appear in 1867.

William Thomson (1824–1907) became professor of natural philosophy at the University of Glasgow, in Scotland, in 1846(!), retiring in 1899, although he remained active in physics until his death. He was knighted in 1865 and became Baron Kelvin of Largs in 1892. His influence on English physics during the reign of Queen Victoria is comparable to that of Helmholtz in Germany.]

4. O. Lodge, *Les Théories modernes de l'électricité. Essai d'une théorie nouvelle*, trans. E. Meylan (Paris: Gauthier-Villars, 1891), p. 3.

5. [Between 1873 and 1878, Thomson redesigned the mariner's compass. His new design compensated for errors introduced by the permanent and temporary magnetization of the surrounding ship. Thomson was best known to the public for

I

If one examines with care the most salient features of English physics, which distinguish it most clearly from French or German science, one soon recognizes that all these features flow from a very deep, very pronounced aspect of the English mind, an aspect that relates some features to others and at the same time explains them.

To a degree one encounters in no other people of Europe, the English possess an imaginative faculty which permits them to represent to themselves a very complex set of concrete things, seeing each in its place, with its motion and its life. On reading a typical English novelist—Dickens, for example—who has not been struck by the abundance and minuteness of the details that overload the least description? To begin with, French readers feel their curiosity piqued by the vivid depiction of each object. But they are unable to see the whole, and the futile effort that they make to reconstruct the innumerable fragments of the picture, scattered before their eyes, soon causes a tiredness that often repels them. The English, however, see the arrangement of all these things without difficulty. Without difficulty their imagination puts each one back in its place, grasps the link that unites them, and finds charming what we find tiring.

This extraordinary power, this abnormal development of the faculty of imagining concrete objects, has its counterpart. Among the English, the faculty of creating abstract concepts, of analyzing them, of relating them by rigorously constructed arguments, seems not to have the strength or the sharpness that the same faculty acquires among Germanic peoples and in our Latin races. English philosophers are almost wholly concerned with applications of philosophy: psychology, ethics, social science. They have little liking for more abstract research and do it poorly. They proceed less by abstract argument than by the accumulation of examples. Instead of connecting deductions, they accumulate facts. Darwin and Spencer do not engage in the learned fencing of discussion with their adversaries; they crush them by stoning them.

This extraordinary power to visualize the concrete, an extreme weakness in grasping the abstract, appears to be the distinguishing feature of

his work on the transatlantic telegraph cable. After providing a mathematical analysis of signal transmission in long undersea cables (1855), he went on to design a series of instruments intended to overcome the difficulties that his analysis had revealed. The most important of these was the mirror galvanometer. The success of the transatlantic cable of 1865 was attributed to his technical expertise; it became the basis for his ennoblement and his large personal fortune.]

the English intellect. It excels at combining things and at creating [fictional] people. It can make the former move and the latter live. But it seems to be unable to give birth to an idea and to develop it. Such appears to be the genius that produced [a] Shakespeare but did not produce a metaphysician.

We will find these two essential traits, these two distinctive marks, again and again while analyzing the form in which the English school has conceived physics.

II

In treatises on physics published in England, one continually finds an element that astonishes French students to a high degree. This element, which almost invariably accompanies the presentation of a theory, is what British scientists call a *model*. Nothing more aptly captures the fashion, so different from our own, in which the English mind proceeds in the construction of science than this use of models.

Two electrified bodies are in evidence. French or German physicists, whether they are called Poisson or Gauss, conceive that, in the space outside these bodies, one places that abstraction called a material point, accompanied by that other abstraction called an electric charge. They then give formulae which permit the determination of the magnitude and direction of the force on this material point when placed at a given geometrical point in the space. Considered at the point in space, the direction of this force touches a certain line: the *line of force*. They demonstrate that the lines of force end at right angles to the surfaces of electrified conductors. They determine the force exerted on each element of such a surface.

This entire theory of electrostatics, formulated in the clear language of analysis and geometry, constitutes a set of ideas and abstract propositions related to one another by rigorous logical rules. This set fully satisfies the intellect of a French or German physicist.

Things go differently for the English. These abstract notions of potential function, equipotential surfaces, and trajectories at right angles to these surfaces fail to satisfy their need to imagine objects that are material, visible, and tangible. "But so long as we adhere to this mode of expression we cannot form a complete mental picture of the actually occurring operations."[6] It is to satisfy this need that they will create a model.

6. Lodge, op. cit., p. 16 [*Modern Views of Electricity* (3rd ed., London: McMillan, 1907), art. 12].

There, where the French or German physicist conceives a family of lines of force, they are going to imagine a bundle of elastic wires. These are stuck by their two extremities to various points on conducting surfaces and stretched; they seek at the same time to shorten themselves and to fatten themselves, to diminish in length and to increase in cross section. When two electrified bodies approach each other [the English] see them drawn together by these wires. Such is the celebrated model of electrostatic actions imagined by Faraday and admired as a work of genius by Maxwell and the entire English school.

The use of similar mechanical models, recalling the essential features of the theory they are trying to present through certain more or less crude analogies, is constant in English treatises on physics. Some, like Maxwell's electrical treatise, make only moderate use of them. Others, on the contrary, make a continuous appeal to these mechanical representations. Here is a book[7] intended to present modern theories of electricity and to outline a new theory. Here there is nothing but ropes running over pulleys, wrapping around drums, running across beads and carrying weights, tubes pumping water, others swelling and contracting themselves, cog-wheels engaging each other and forming pinions for racks.

It is far from the case that these models help French readers to understand a theory; on the contrary, in many cases the French must make a serious effort to understand the functioning of the apparatus that the author has described, which is sometimes very complicated. This effort is often much greater than that required to understand the abstract theory, which the model claims to embody in its pure form.

Yet the English find the use of a model so necessary for the study of physics that, for them, designing the model is mistaken for understanding the theory itself. It is amusing to see this confusion formally accepted by the very person who is the highest expression of English genius today, W. Thomson:[8]

> It seems to me, he says, that the test of "Do we or not understand a particular subject in physics?" is, "Can we make a mechanical model of it?" I have an

7. Lodge, op. cit.

8. [*Notes of Lectures on Molecular Dynamics and the Wave Theory of Light.* Delivered at *The Johns Hopkins University, Baltimore, by Sir William Thomson, Professor in the University of Glasgow. Stenographically Reported by A. S. Hathaway, Lately Fellow in Mathematics of The Johns Hopkins University.* Baltimore: Johns Hopkins, 1884. (Reproduced by the "papyrograph" process.) The passages quoted are from, respectively, p. 132 and pp. 270–271.]

immense admiration for Maxwell's model of electromagnetic induction. He makes a model that does all the wonderful things that electricity does in inducing currents, etc., and there can be no doubt that a mechanical model of that kind is immensely instructive and is a step towards a definite mechanical theory of electromagnetism. . . .

I never satisfy myself until I can make a mechanical model of a thing. If I can make a mechanical model I can understand it. As long as I cannot make a mechanical model all the way through, I cannot understand, and that is why I cannot get *the* electromagnetic theory. I firmly believe in *an* electromagnetic theory of light, and that when we understand electricity, magnetism and light, we shall see them all together as part of a whole. But I want to understand light as well as I can without introducing things that we understand even less of. This is why I address myself to pure dynamics.[9]

III

For physicists of the English school, understanding a physical phenomenon is the same thing as constructing a model that imitates the phenomenon. Consequently, understanding the nature of material things will be the same thing as imagining a mechanism that will represent or simulate the properties of bodies by its action. The English school has thus acceded entirely to purely mechanical explanations of physical phenomena.

This is not, to be sure, a characteristic that suffices to distinguish English doctrines from scientific traditions that flourish in other countries. Mechanical theories have resulted from French genius, the genius of Descartes. They have long reigned without dispute in France as in Germany. What distinguishes the English school is not the attempt to reduce matter to mechanism, it is the particular form of its attempts to reach this goal.

No doubt, wherever mechanical theories have taken root, wherever they have been developed, they have owed their birth and progress to a failure of the faculty of abstraction, to a victory of imagination over reason. If Descartes and the philosophers who followed him refused to admit the existence of any property of matter not reducible to geometry or kinemat-

9. Pp. 270–271, emphasis added by Duhem. [The last sentence is a translation of Duhem's French. The English text (Thomson, *Molecular Dynamics and the Wave Theory of Light*, p. 271) reads, "That is why I take plain dynamics." Continuing, "I can get a model in plain dynamics, I cannot in electromagnetics." "Plain dynamics" here means dynamics limited to the fundamental concepts and relations of Newtonian mechanics, and not augmented by additional concepts or relations from electricity, magnetism, and so forth. Understood in this way, Duhem's translation of the final sentence makes good sense and provides a better ending.]

ics, it is because any such quality would be occult, and, being conceivable only by reason, it would remain inaccessible to the imagination. The reduction of matter to extension by the great thinkers of the seventeenth century showed clearly that during that period, the metaphysical sense, exhausted by the excesses of scholasticism during its decadence, entered into the decrepit state in which it still languishes today.

But in France, as in Germany, while the sense of abstraction could have failures, it never completely went to sleep. It is true [that] the hypothesis that everything in the material world can be reduced to geometry and kinematics is a triumph of imagination over reason. But after having given in on the essential point, reason, at least, reasserts its rights when it attempts to deduce the consequences and to construct the mechanism which must represent matter. Descartes, for example, and Huygens after him, having posited the principle that extension is the essence of matter, took great care to deduce from it that matter has the same nature everywhere, that there cannot be several different material substances, and that only forms and motions could distinguish the different parts of matter from one another. By logic, they sought to construct a system that explains natural phenomena through the intervention of only two elements: the shape of the bare parts and the motion with which they are animated.

In addition, since the faculties of the French and the Germans do not permit them to imagine a mechanism when it is at all complicated, the French and the Germans demand that all attempts to explain the universe mechanically be *simple*. Any explanation that made a considerable number of elementary substances intervene, or that combined them into a complicated organism, would be rejected by them as improbable from the outset. They [would] demand that one reduce matter, in the last analysis, to a small number of types of elementary atoms, two or three at most; that these atoms have simple geometric forms; that they be endowed only with some essential mechanical properties; and that the properties be set forth in propositions that are very brief and easy to understand—propositions which they would seek, moreover, to justify through metaphysical considerations. Let one examine all the mechanical explanations imagined by the French or the Germans from the doctrine of Descartes right up to the theories of P. Leray, which we analyzed here a short time ago,[10] and one will always recognize there, very clearly, the dual tendency toward abstraction and simplicity.

10. P. Duhem, "Une nouvelle théorie du monde inorganique," *Revue des questions scientifiques*, Jan. 1893.

Things are not the same with the mechanical explanations created by British intellects. Their powerful imaginative faculty depicts the most obscure mechanisms for them without difficulty. Also, they are not afraid of attributing a very complex structure to matter. In order to explain the dispersion of colored light, W. Thomson considers material molecules to be veritable edifices with interacting rigid and elastic elements. His gyrostatic aether is not very simple, and yet it greatly surpasses in simplicity the aether that Maxwell and Oliver Lodge constructed to give an account of electromagnetic phenomena.

It is not only that the constructions through which English physicists seek to represent the constitution of matter are complicated, but also their constituents do not reduce to geometric forms endowed with a few elementary abstract properties. These are not the materials with which Descartes sought to construct the "machine" of the world, simple shapes endowed with the property of exchanging their quantity of motion through collisions without losing any of it. No, these are concrete objects, similar to those we see or touch: rigid or elastic solids, compressible or incompressible fluids. At times, in order to make them more tangible, in order to make us better understand that this is not a case of ideas developed by abstraction, but rather of bodies equivalent to those we manipulate every day, W. Thomson affects to designate them with ordinary names. He calls them little wires or bell-pulls. The elementary properties with which these bodies are endowed—rigidity, elasticity, compressibility, fluidity, and flexibility—receive neither definition nor metaphysical justification. W. Thomson, for example, never asks himself philosophical questions such as the following: Can one of the ultimate elements of matter occupy a variable volume or not? Is it essentially incompressible, or can it be compressed? Still less does he ask himself what one must understand by the volume occupied by an atom. The elements which make up matter are similar to those which we see around us every day. They may be fluid like water, compressible like air, elastic like steel, or flexible like a strand of silk. Their nature does not need to be defined philosophically. It suffices that their properties fall under the senses. The mechanisms they serve to make up are not destined to be grasped by reason; they are destined to be seen by the imagination.

IV

What we have said about the use of "models" to "illustrate" physical theories will help us understand the role that the English assigned to mathematics in the development of the same theories.

Certainly, more than one reader may be astonished to hear us speak of the role that accrues to the imaginative faculty in mathematical research. Mathematics passes as a science that only the faculty of creating abstract ideas, joined to the faculty of connecting them in logical reasoning, has the power to engender and to develop. This common opinion, however, does not seem to me to be completely right, at least unless it is explained.

Without doubt, all branches of pure and applied mathematics treat concepts that are abstract. It is abstraction that furnishes the notions of number, line, surface, angle, mass, force, temperature, and quantity of heat or electricity. It is abstraction, or philosophical analysis, that separates and makes precise the fundamental properties of these various notions and enunciates axioms and postulates. It would be possible to connect these abstract notions through reasoning that would involve almost exclusively the logical faculties of the mind. Euclidean geometry offers us an example of a similar connection. Mathematical procedures have precisely as a goal to replace this extremely laborious method with another which is much easier. Instead of reasoning directly from the abstract notions that concern it, considered in themselves, the mathematician profits from their most simple properties in order to represent them by numbers, that is, to *measure* them. Then, instead of connecting the properties of these notions themselves in a chain of syllogisms, the mathematician submits their measures to manipulations according to fixed rules, the rules of mathematical analysis. For in mathematical analysis a very important part, which one can in the widest sense of the word designate as *calculation*, presupposes, among those who developed it or who use it, a good deal less a power of abstraction and the facility of setting their thoughts in order than the ability to represent to themselves various and complicated combinations that one can form with certain symbols; that is, to see the transformations which allow one of these combinations to pass into another. The authors of certain analytic inquiries are not at all metaphysicians. They resemble the engineer who combines multiple wheels, or even better, the chess player who brings out a bishop and a knight without looking at the board.

After what we have said about the English intellect, one might think that the geometers of Great Britain would excel at manipulating the most complex algorithms in algebra, just as much as in deepening the very principles on which mathematics rest. Does this expectation not find itself confirmed in a resounding manner if one compares the researches of someone like Sylvester to those of someone like Riemann or Weierstrass?

For the English, mathematics consists above all of an algebraic mechanism. What role can it play in the development of physical theory? That of a *model*. Just as, in order to clarify a physical theory, they constructed an apparatus with materials that are solid or liquid, elastic or flexible, the action of which imitated the principal phenomena that the theory aimed to capture, similarly, with the symbols of algebra, they aim to construct a system representing the coordination of the laws that they seek to classify through its various transformations. When they construct a model, they make it from the materials which seem to them to be the most convenient without ever asking themselves if the arrangements they imagine have the least analogy in nature with the properties of the bodies that they want to reproduce. The same is true even when it is a case of representing the constitution of matter. Similarly, when they construct a mathematical theory, they are little concerned to understand what real elements correspond to the algebraic magnitudes that appear in their equations. If these equations imitate the interplay of the phenomena well, they are little concerned with the route by which they were obtained.

Those who founded mathematical physics in France or Germany— Laplace, Poisson, Ampère, Gauss—took great care to note the facts of experience on which they founded it when introducing a physical theory, to state the hypotheses they admitted precisely, [and] to define the magnitudes of which they spoke. In this way, these preliminaries, in general so carefully attended to, led the reader step by step right up to an equation in which the complete theory is condensed. Almost always one looks in vain for these preliminaries among English authors. For them the equation alone has value. The background of the equation has no interest for them.

Would you like a striking example of this?

Maxwell has added a new electrodynamics, the electrodynamics of dielectric bodies, to the electrodynamics of conducting bodies created by Ampère. This branch of physics results from the consideration of an essentially new element, which has been named, improperly in any case, *displacement current*. [The displacement current is] introduced to complete the definition of the variable state of a dielectric, a state which the knowledge of polarization does not determine entirely, just as *conduction current* has been added to electric charge to complete the definition of the variable state of a conductor. The displacement current presents strict analogies as well as profound differences with the conduction current. Thanks to the introduction of this new element, electrodynamics is thrown into disorder. Phenomena are announced that experimentation has never revealed. We see a new theory develop about the propagation of electric actions in non-conducting media, and this theory leads to an unforeseen interpretation of optical phenomena. The displacement current is so new, so strange. Its study is so fertile in consequences that are important, surprising, and paradoxical. Surely Maxwell would only introduce this new element into his equations after having defined and analyzed it with the most minute care? Open the work[11] in which Maxwell presented his new theory of the electromagnetic field and you will find nothing but these two lines to justify introducing the flux of displacement in the electrodynamic equations: "The variations of electric displacement must be added to the currents in order to obtain the total motion of electricity."

This absence of all definition, even when we are concerned with the most novel and important elements, allows us to understand how some people for whom analysis holds no mystery remain excluded from the work of Maxwell and become incapable, in many cases, of saying what he really thinks. Maxwell studies the transformation of the equations of electrodynamics in their own terms, most often without seeking to see behind his transformations the coordination of physical laws. He studies them as one examines the movements of a mechanism. This is why it is a futile effort to seek behind these equations a philosophical idea which is not there. It seems to me that this is the sense in which one must interpret the remark of Hermann Hertz:[12] "To the question 'What is Maxwell's theory?' I know no more complete and simpler answer than this: Maxwell's theory is Maxwell's system of equations."

V

The French geometers who constructed the most important theories in mathematical physics had a continual tendency to regard them as true explanations, in the metaphysical sense of the word. They assumed that these theories took hold of the very reality of things and the true causes of phenomena. This tendency, springing from Descartes, constantly reveals itself in the writings of Laplace, Poisson, Fresnel, Cauchy, and Ampère. Occasionally, it is true, these authors are almost afraid of their own bold-

11. J. Clerk Maxwell, "A Dynamical Theory of the Electromagnetic Field," *Philosophical Transactions of the Royal Society* 155: 480.

12. Hermann Hertz, *Untersuchungen über die Ausbreitung der elektrischen Kraft*, Einleitende Übersicht, p. 23. Leipzig 1892. [In English, the first physicist to detect radio waves experimentally, Heinrich Rudolph Hermann Hertz, is generally known as Heinrich Hertz.]

ness, and for an instant, they suspect that their theories are perhaps no more than representations and not explanations at all.[13] But after occurring to them for a moment, and revealing to these great minds the true import of the method in use in the positive sciences, this prudent thought is veiled anew and disappears behind the clouds gathered by a superb and absolute confidence in the omnipotence of modern science.

This tendency to see a metaphysical explanation of the universe in mathematical theory is in singular contrast with the tendency of English physicists, who never see it as anything but a model. Even when he is writing a paper on the constitution of the aether or of matter, W. Thomson never forgets that he is not laying hold of the essence of things. He confines himself to constructing an apparatus capable of simulating certain phenomena. This thought is always present in his mind. He returns to it every instant.

This opposition between the French and English tendencies shows itself in essential and striking features.

At the beginning of every theory, French physicists of the late eighteenth and early nineteenth century present a certain number of hypotheses which define the most important, essential, and elementary properties of matter for them. Then, from these fundamental hypotheses, they seek to deduce the explanation of all the phenomena of physics through a logically connected chain of precise deductions. Nothing can remain outside the chain, for the fundamental hypotheses are supposed to define all the primary properties of matter from which flow all the phenomena that we observe, as effects flow from causes. This method has produced those majestic systems of nature which propose to bring the form of Euclidean geometry to physics. Taking as foundations a certain number of very simple postulates, they aspire to deduce from them the explanation of the material world, down to the last detail. From the time when Descartes unrolled the ample chain [of deductions] in his *Principles of Philosophy* right up to the time when Poisson, following Laplace, sought to reduce the mechanism of the system of the world to attraction, as much Newtonian as molecular, and thus to construct all of *Mechanical Physics*, such has been the perpetual ideal of French intellects. In pursuing this ideal, monuments

13. [As this passage suggests, for Duhem there is a profound difference between representation and explanation. The latter, not the former, must lead back to the one correct ontology for science, that provided by the "natural classification" that is expected to be the historical end point of physical theorizing. See Duhem, trans. Weiner (1954), esp. pp. 296–297.]

have been erected whose grandiose proportions and simple lines still enchant, though today they totter on foundations undermined on all sides.

This unity of theory and the logical chaining together of all the constitutive parts are such necessary and compulsory consequences of the manner in which physicists of the French school conceive a theory that to disturb this unity or to break this order is, for them, to violate the principles of logic. It is to commit an *absurdity*.

It is not at all the same for English physicists.

For them, a mathematical theory is not an explanation of physical laws but a model of these laws. [Such a theory] is constructed not to satisfy reason but to please the imagination. Henceforth, mathematical theory escapes the domination of logic. English physicists are permitted to construct one model to represent one group of laws and another model, unconnected with the former, to represent another group of laws, even when some laws are common to the two groups. For a geometer of the school of Laplace and Cauchy, it would be absurd to give two distinct explanations of one law and to maintain that these two explanations are true at the same time. For an English physicist, there is no contradiction in one law being represented in two different ways by two different models. There is more: The complication thus introduced into science causes no offense, for [the English] imaginative faculty, which is more powerful than our own, does not feel the desire for simplicity and the need for unity to the same degree that we do. This faculty finds its way without difficulty in labyrinths where ours gets lost.

Hence, in the theories of the English, we are compelled to judge severely these disparities, or incoherencies, or contradictions, because we seek a rational system in which the author aspires to give us nothing but a work of the imagination.

In reading these lectures of W. Thomson in *The Constitution of Matter*, take good care to seek in them a set of logically coordinated researches showing how different physical laws may be deduced from chosen hypotheses on the constitution of matter. Great will be your surprise and still greater yet will be your disappointment. Here matter is presented to us as a set of isolated and immobile material points. Between these points attractions exist, and W. Thomson, having announced the idea that the attractions can be reduced to Newtonian action, develops the hypothesis which distinguishes them [from Newtonian actions]. There, gases are a collection of tiny projectiles, animated with prodigious velocities, which collide in their mad progress. Further on, a material molecule is a collection of spherical concentric envelopes connected by springs. Elsewhere, it is a gyrostatic system constituted by whirlpools in the aether. There is no attempt at agreement between these various theories. Each of them is developed in isolation, with no concern for the preceding one, covering again a part of the field already covered by the preceding model. They are pictures, and in composing each one, the artist chooses the objects which he will represent and the order in which he will group them, with complete freedom. It matters little if one of his figures has already appeared, in a different pose, for another picture. The logician would be wrong to take offense at this. A series of pictures is not a chain of syllogisms.

This incoherence between the various parts of a theory is not limited to W. Thomson. It is yet more striking in the writings of Maxwell. As H. Poincaré says in an already famous preface:[14]

> The English scientist does not seek to build a unique, definitive and well ordered structure. It seems instead that he raises a great number of independent, provisional constructions, between which communications are difficult and at times impossible.
>
> Let us take, as an example, the chapter explaining electrostatic attractions by compressions and tensions which might hold sway in a dielectric medium. This chapter could be eliminated without the remainder of the book becoming less clear and complete. On the other hand, it contains a theory that is sufficient unto itself. One could understand it without having read a single line before or after. But it is not just independent of the rest of the work: It is difficult to reconcile with the fundamental ideas of the book, as an extensive discussion will show. Maxwell himself does not even attempt such a reconciliation. He limits himself to saying, "I have not been able to make the next step, namely, to account by mechanical considerations for these stresses in the dielectric."[15]
>
> This example will suffice to make my idea understood. I could cite many others. Hence, who would doubt, on reading the pages devoted to the magneto-optical rotation,[16] that an identity exists between optical and magnetic phenomena?

14. H. Poincaré, *Electricité et optique* I. *Les Théories de Maxwell et la théorie électromagnetique de la lumière*, Introduction, p. viii.

15. [James Clerk Maxwell, *A Treatise on Electricity and Magnetism*, 3rd ed., New York: Dover, [1891] 1954, vol. 1, art. 111, p. 166.]

16. ["la polarisation rotatoire magnetique." Faraday discovered that when a beam of light is passed through a glass plate in a strong magnetic field, its plane of polarization is rotated. This effect, known in English as "the magneto-optical rotation," was taken by Maxwell and his followers as evidence not only that the domains of magnetism and light were connected but also that magnetism consisted most

No doubt what is exact and truly fertile in the work of Maxwell will one day take its place in a coherent and logically constructed system, in one of those systems in which thoughts are conducted in order, in the image of Euclid's *Elements*, or of those majestic theories unfolded by the creators of mathematical physics. But Maxwell most assuredly was not seeking that. For example, regarding Boltzmann's[17] attempt to construct an equivalent system, we ought to see in his attempt not the work of a commentator scrupulously and slavishly faithful to the great physicist, but the work of a German thinker seeking to transform into a logically coordinated combination of rational theories what, in the mind of the English author, was nothing but a set of models constructed for the imagination.

When one traverses the works of a great English physicist such as W. Thomson or Maxwell, when one sees the appearance of these disparate views, contradicting themselves from one year of his life to another and from one chapter of his book to another, one takes to musing on those innumerable laws and customs that each century adds to English legislation—laws and customs that contradict the customs and laws of preceding centuries and nevertheless, far from destroying [their predecessors], superimpose themselves on them, mixing and confusing themselves with them. It is striking to find again, in science just as in law, this careless logic before which the French mind, thirsting for simplicity and unity, is stupefied. In all manner of things, the French demand a system.

VI

Here [is] a digression on a subject that seems important to us.

The geometers, French for the most part, who founded mathematical physics, saw the rational explanation, or the metaphysical foundation, of the laws discovered by experimentalists in the theories constituting that science. From then on, they wanted these theories to be logically connected.

Today, this way of understanding the role of theories in mathematical physics is in the process of being abandoned. Physicists, more and more, at least those who reflect on the meaning of the science they are obliged to develop and teach, no longer tend to see in physical theories metaphysical

---

importantly of a rotational motion in the luminiferous aether. See Maxwell, *Treatise*, vol. 2, arts. 806–831.]

17. Boltzmann, *Vorlesungen über Maxwell's Theorie der Elektricität und des Lichtes*, Teil 1, Leipzig, 1891.

explanations but only representational systems that classify and coordinate physical laws. We have developed the reasons that compel the adoption of this idea, in our view, in several earlier papers in this journal.[18]

Now, if we admit that the theories of mathematical physics are not metaphysical systems, if we attribute to them only a value as representations, if we consider them as methods of classification, why should we still require that all these theories be derived with absolute rigor from a small number of principles stated clearly and once and for all? Why should we not admit that distinct groups of laws may be symbolized by different theories, the first resting on certain hypotheses, the others on other hypotheses incompatible with the first? Similarly, why should we not admit several different and irreconcilable theories at one time [in order] to give an account of one of the same collection of phenomena. Why, in a word, should we prefer the logical rigor of French theorists to the logical incoherence of English physicists?

This thought certainly presented itself to quite a few minds.

No doubt, there are some for whom this idea leads to skepticism. They are not far from placing the method followed by Laplace and Ampère on the same level as the method followed by Thomson and Maxwell. Perhaps they are even inclined to prefer the latter. Is this not the tendency which comes through in the following lines, written by H. Poincaré?[19]

> One should not, therefore, expect to avoid all contradictions, but one should take sides. Two contradictory theories may in fact each be useful instruments of research, provided that one does not mix them, and that one does not look there for the essence of things. Perhaps Maxwell's work would be less suggestive if he had not opened for us so many new and divergent paths?

Others, on the contrary, those who wish absolutely to attribute an ontological value to physical theories—Vicaire,[20] for example—are content to show that in regarding physical theories as pure representations, one is led to regard logical incoherence in equivalent theories as legitimate, and they

18. [Chapters 1 and 2 of this volume]; "Notation atomique et hypothèses atomistiques," *Revue des questions scientifiques* (1892); "Une nouvelle théorie du monde inorganique," *Revue des questions scientifiques* (1893).

19. H. Poincaré, *Électricité et optique*. I. *Les Théories de Maxwell et la théorie électromagnétique de la lumière*, Introduction, p. ix.

20. E. Vicaire, "De la valeur objective des hypothèses physiques," *Revue des questions scientifiques* (1893).

are well aware that this consequence, which inspires such violent loathing among French minds, comes back as an objection against the ideas from which it came.

The ideas in the process of being born and evolving among physicists therefore pose an important problem. This problem may be stated as follows: *Is logical incoherence legitimate in theoretical physics?* Or again, in a more explicit way, in the following manner: *Is it legitimate to formulate several distinct groups of experimental laws, or even a single group of laws, by means of several theories, each of which rests on hypotheses irreconcilable with those that support the others?*

To this question, we do not hesitate to respond, as we have already:[21] IF WE RESTRICT OURSELVES TO INVOKING CONSIDERATIONS OF PURE LOGIC, *we cannot prevent a physicist from representing different sets of laws, or even a single group of laws, by several irreconcilable theories. One cannot condemn incoherence in the development of physical theory.*

In fact, if we admit, as we have sought to establish, that a physical theory is nothing but a classification of a group of experimental laws, how can we extract, from the laws of logic, the right to condemn a physicist who employs different procedures of classification to coordinate different groups of laws, or who proposes differing classifications resulting from different methods for one group of laws? Does logic forbid naturalists to classify one group of animals according to the structure of the nervous system and another group according to the structure of the circulatory system? Would a malacologist be absurd to simultaneously set forth the classification of Bouvier, who classified mollusks by the arrangement of their network of nerves, and that of Remy Perrier, who based his comparisons on the study of the organ of Bojanus?[22] Thus, logically, the physicist should have the right to regard matter as continuous in one place and envision it as formed from separate atoms in another; to explain capillary action by means of attractive forces acting between stationary particles, and to endow these same articles with rapid motion to take account of the effects of heat. None of these disparities violates the principles of logic.

Clearly, logic imposes only a single obligation on physicists, and that is not to mix their different procedures of classification. When physicists establish a certain relationship between two laws [they are obliged] to say precisely which among the methods they use justifies this relationship. In

21. [Chapter 1 of this volume], section 8.

22. [The excretory organ of a lamellibranch (named for the Alsatian zoologist L. H. Bojanus).]

a word, [physicists are obliged] as H. Poincaré put it, not to mix two contradictory theories.

If we invoke exclusively considerations from the domain of logic, we are therefore unable to condemn logical incoherence in physical theories. But considerations of pure logic are not the only ones that reasonably direct our judgment. The principle of contradiction is not the only one to which we would be permitted to have recourse. In order that we should legitimately reject a method, it is not necessary that [the method] be absurd. It suffices that our aim, in rejecting it, is to prefer a better method. It is by virtue of this principle that we are able to solve the difficulty which we are examining and to offer the following rule legitimately: *In physical theory, we must avoid logical incoherence* BECAUSE IT INJURES THE PERFECTION OF SCIENCE.

It is better, it is more perfect, to coordinate a set of experimental laws in the midst of a single theory, where all the logically connected parts follow in undeniable order from a certain number of fundamental hypotheses stated once and for all, than to invoke, in classifying these same laws, a great number of irreconcilable theories, some of which rest on certain hypotheses [and] others of which rest on other hypotheses contradicting them. It is a truism that everyone admits without needing to remark on it that even those such as the English physicists or their imitators, who most gladly accept contradictory theories to take account of different laws, nevertheless prefer a unique theory when they easily see the means to construct it. This truism provides us with an example of the clear and self-evident principles on which the application of the experimental method depends, as I have indicated elsewhere.[23]

But although this truism is so clear and so evident that all physicists make use of it without hesitation in the course of their researches, it does not follow from it that metaphysicians do not have to take it into account; not, certainly, in order to increase clarity, which is complete, or certainty, which is intuitive, but in order to make us take hold of the relations of this principle with other principles that guide our reason and in order to disarm skepticism when it has thought to undermine the foundation of physical theory.

Why, therefore, is a coherent physical theory more perfect than an incoherent collection of incompatible theories, even in the eyes of those who do not value physical theories as metaphysical explanations?

We must evidently judge the degree of perfection of a physical theory by the greater or lesser conformity which that theory offers to the ideal and

23. [Chapter 2 of this volume], section 3.

perfect theory. Now we have defined this ideal and perfect theory elsewhere. It would be the complete and adequate metaphysical explanation of material things. This theory, in fact, would classify physical laws in an order which would be the very expression of the metaphysical relations that the essences that cause the laws have among themselves. It would give us, in the true sense of the word, a *natural classification* of laws.

Such a theory, like everything that is perfect, infinitely surpasses the scope of the human mind. The theories which our methods permit us to construct are no more than a pale reflection of it. The metaphysical method gives us only information that is too general, too lacking in detail, and too paltry about the essence of material things to be able to serve in classifying physical laws. The experimental method, the only one to which we are able to have recourse in pursuit of this goal, does not capture the essence of things, but only the phenomena through which things manifest themselves to us. It does not allow us to reconcile the laws with one another except through exterior and superficial analogies which translate the true affinities of the essences from which the laws emanate [and] perhaps frequently betray them.

But however imperfect physical theories are, they can and they should tend toward perfection. No doubt they will never be anything but a classification, stating analogies between laws but not capturing the relations between essences. We can and should always seek to establish them in such a way that there would be some probability that the analogies brought to light by them would be not accidental agreement, but true relations, showing the connections that really exist among essences. In other words, we can and should seek to render these classifications as far from artificial, as *natural* as possible.

But if we know little about the relations which material substances have among themselves, there are at least two truths of which we are certain: These relations are neither indeterminate nor contradictory; therefore, whenever physics presents us with two irreconcilable theories about one group of laws, or, again, whenever [physics] symbolizes a group of laws by means of certain hypotheses and another group of laws by means of other hypotheses that are incompatible with the preceding ones, we are certain that the classification which such a physics presents to us does not conform to the natural order of the laws, [that is] to the order in which an intelligence that sees essences would arrange them. By making theoretical incoherences disappear, we would have some chance to approach this order, to make [theory] more natural and thus more perfect.

VII

Let us return to the study of the characteristics that distinguish physicists of the English school.

The need to connect their deductions in a logical chain and to conduct their thoughts in an orderly way obliges French or German physicists to be prudent and even faint-hearted. They will not tolerate either contradictions or gaps in their theories. Consequently, it seems to them that any proposition not clearly and evidently connected with accepted principles, any that is strange, any that is surprising, must be called in question for that very reason.

Matters are completely different with the English. Strange things do not frighten them. Surprises do not create doubt for them. They seem, on the contrary, to seek out all that is unforeseen and all that is audacious in the domain of science.

Whereas French physicists and, above all, German physicists, when they have discovered a new law, like to relate it to accepted principles, to show that [the law] follows from them naturally, the English, on the contrary, take pleasure in giving a paradoxical form even to the logical consequences of the most universally accepted theories. This tendency is apparent in the various applications that W. Thomson has made of the principles of thermodynamics, and it appears above all very clearly when one compares the papers that he has devoted to these questions with those written on the same subjects by Clausius.

The prudent mind of continental physicists is characterized above all by their hesitation in engaging certain questions situated on the borders of science: the innermost constitution of the material world, what existed millions of years ago, what will exist in millions of years. We cannot see these questions, so vast, so complex, so troubling, resolved without a shiver of skepticism making us tremble. The English ignore these fears. The size and dimension of atoms, the constitution of matter, the nature of light and electricity, the dissipation of energy, the origin and duration of solar heat—here are the problems that attract W. Thomson, Maxwell, [and] Tait. Their vigorous imagination deploys itself there with ease in audacious leaps that make no obstacle of the bounds of logical rigor. [Their imagination] enjoys playing with numbers that are terrifying because of their large or small size, just as athletes enjoy the prodigious exercises that make them aware of their muscles' power.

In the leaders of the English school, in William Thomson, in Maxwell, this tendency to treat strange and troubling things knows some limits. It has no limits in their disciples. William Crookes, Oliver Lodge, and Tait treat convulsions of the modern imagination that reason no longer holds in equilibrium, such as communication of thoughts at a distance, spiritualism, and magic, with the same confidence, the same tranquillity, with which they treat a question in optics or electricity. For them the bizarre has every chance of being true.

This boldness of the English mind presents great dangers for a science that is no longer on guard against extravagance. On the other hand, it has advantages. It favors invention to a high degree.

Our need to admit nothing except what can be clearly deduced from accepted principles makes us mistrustful of any unexpected discovery. This need leads to the bureaucratic mind, hostile to novelties, for which continental scientists and their academies are so often reproached. Inventors find not only around them but also within them this fear of the unexpected, which is a born enemy of inventive genius. Their intellect itself refuses to admit the exactness of the new idea that grows within them to the extent that their intellectual faculties have not analyzed that idea and have not made it enter into a system of logically connected deductions. One can therefore understand that inventions hatched on the continent may be neither as numerous nor, above all, as audacious as the inventions born in England or America. Inventors in England or America would not be held back by the same difficulties, or exposed to the same hostilities, as those in France or Germany.

In England, inventors find conditions within and around themselves that assure free development and a favorable reception for their ideas. The same is true for Thomson, our lecturer.

Among those who have studied science little, the imagination primes the reasoning faculties. The solidity of principles, the rigor of deductions, interest them less than do the boldness and strangeness of conclusions. Lecturers must therefore address themselves to the imagination of their audience and not to its reasoning faculty. This is what makes French scientists unsuitable in the role of lecturer. They cannot bring themselves to state propositions with no logical connection, and when their audience is not in a condition to understand the real connection which unites these propositions, they prefer to establish false and artificial connections rather than to establish none. Besides, they are constrained by the audience itself, which demands that they prove everything which has been stated to them, that everything is explained to them that has been shown to them, even though they would be incapable of understanding the proofs and following the explanations. From this [follows] the lack of sincerity and the air of charlatanry in the proceedings that lectures readily take on among the French. From this [follows] the scorn that the majority of serious scientists direct toward this form of teaching.

English scientists, on the contrary, have great affinities with their audience. As is true of the audience, their imaginative faculty is more developed than is their deductive faculty. They feel no need to chain syllogisms together. The field their intellect enjoys [is] that of facts, abundant facts, living facts, facts complicated if they must be, but as far as possible strange facts, unforeseen facts. This also is what is most likely to captivate their audience, which demands to see rather than to understand. Thus is explained the success of the essays of Tait, of W. Thomson, and the prodigious triumph of Tyndall's lectures.

VIII

I know no matter more appropriate for letting us take hold of the characteristics of English science than the comparison of the works of Thomson with those of Helmholtz. Connections abound between these two geniuses. [They share] the same precociousness, making the strokes of their writing the strokes of a master. [They possess] the same fertility, which forty-five years of continuous scientific production has not exhausted. [They display] the same breadth of thought, which embraces the most diverse subjects without difficulty and treats them with equal originality. [They share] the same renown, which the countrymen of these two scientists cite with pride, princes recognize with noble titles, and all Europe greets with approbation. And meanwhile, for those who reflect while reading their works, what differences, what contrasts between Thomson and Helmholtz! The one is the English mind, in all its fullness; the other the German mind.

What is striking, at first approach, in Helmholtz's work is the logical power which gives to the work so majestic a unity and so broad an application. The broad lines of that work may be traced from the first work of Helmholtz, from the monograph *On the Conservation of Force*, which would be to science like a manifesto from which a revolution bursts forth. Then, with an attention to detail the like of which science offers few examples, Helmholtz took up again each of the subjects that he had outlined, made its outline more precise, enlarged it, deepened it, and from something which seemed to be only a remark, he made a whole branch of science grow. Follow the development of a single one of these topics, electrodynamics. At the outset, in *Die Erhaltung der Kraft* (*The Conserva-*

tion of Force), only a few pages are devoted to it. Here are the seeds sown by Helmholtz: the central idea of the electrodynamic potential, and a view on the relations between electrodynamic actions and the principle of energy conservation. Consider now the tree as it develops. The idea of electrodynamic potential has become the strong trunk, from which spread out like branches the theory of induction, the laws of force between currents, the properties of dielectric bodies and of magnetic bodies. The insight on the link that the principle of energy conservation establishes between ponderomotive and electromotive forces has engendered these fundamental articles in *On the Role of the Principle of Least Action in Physics*, which reunites electrodynamics with mechanics, thermodynamics, and optics. Thus, it rises up, like a mighty oak, this synthesis which seems to have absorbed, elaborated, and made fruitful anything that had a spark of life in the electromagnetic works of W. Weber, F. E. Neumann, Maxwell, Kirchhoff, and C. Neumann.

But the whole genius of Helmholtz is not only a power of general application that extends or a force of logic which leads everything back to unity. To these two qualities, which he possesses to an eminent degree, must be joined still a third: penetrating analysis that dissects and reduces the questions submitted to him to their ultimate elements and their irreducible principles. From there [flow] his deep researches on the foundations of geometry and his meditations, so satisfying to the mind, on the origins of the axioms of arithmetic. It is this power of analysis that explains the unity and breadth of Helmholtz's theories. If the oak is unshakable, if its branches are robust, if its dense foliage shades a vast prairie, it is because its roots penetrate deeply into the soil, assuring a firm point of purchase and an abundance of nourishing sap.

What a contrast with the work of Thomson! The unity, the generality of application, the depth of Helmholtz's theories have disappeared to make room for an infinite variety of brilliant, ingenious, sometimes inspired views, each of which develops for its own ends and with no regard for the others. It is no longer an oak we have before our eyes; it is a spray of flowers in a thousand shapes, a thousand colors, the stems of which entangle without supporting one another. Prudently, Helmholtz ceaselessly surveys the ground on which he builds and assures the firmness of his building's foundations. Thomson, less concerned with rigorous principles, goes right to the most remote consequences, [those that are] the most bold, sometimes the most reckless and the most risky. Departing from physics, Helmholtz ascends through analysis, from principle to principle, to the verge of metaphysics. Thomson descends, from consequence to consequence, to the verge of industrial applications. The former is one of the most profound philosophers of our century; the latter is one of its most inventive engineers.

IX

When one asserts, before certain persons, that there is an *English* manner of conceiving physical science, which is very different from the French manner or from the German manner, one sees that they are astonished. Is not science essentially international? Poincaré pictures for us the surprise of a French reader opening Maxwell's treatise. "What does he understand a French reader to be?" exclaims Joseph Bertrand.[24] "Why suppose that the English or the Germans would be less shocked by the lack of rigor? Did two centuries suffice to change the national intellect? Do Newton's descendants today accept imagination in physics and leave to Descartes's compatriots respect for rigor and love of precision?"

It is beyond argument that logic is unitary. Its principles impose themselves, with the same ineluctable rigor, on the French, the English, and the Germans. The prohibitions pronounced by logic extend to all lands, and no asylum can protect those who incur them. But if the laws of logic are the same at all times and in all countries, if everywhere and always those who respect logic are obliged to reason in the same manner, there are, by way of return, infinite ways to disobey logic or to sin against it. These violations of the laws of logic submit to the influence of the epoch and the context in which they are committed. Truth, being impersonal, carries no sign of the circumstances in which it was discovered. Error, being human work, results from human habits, from human prejudices, from ideas that surround humans, from ignorance of the context in the midst of which humans live. Error varies with these conditions and is explained by them. In the same way, the moral law is the same on either side of the Pyrénées. But do violations of this law, do immoral acts taken together, present the same general aspects in France and in Spain? Do they not submit to the influence of [different] peoples and contexts?

In the domain of science it is not error alone that carries the special mark of the people among whom a doctrine germinated and grew. We have often said, and cannot too often repeat, [that] theoretical research is not submitted whole to the inflexible laws of logic, in each of its parts, [or] in each of the operations which make it up. Certain of the elementary opera-

24. Joseph Bertrand, *Journal des savants*, Dec. 1891, p. 743.

tions that constitute it—for example, the choice of the hypotheses on which each theory rests—escape in many ways from the scope of these laws. There, where logic fails to trace a path for physicists from which they cannot stray, the special cast of their mind, their dominant [mental] faculties, the doctrines widespread among their peers, the traditions of their predecessors, the habits they have formed, and the education they have received [all] serve them as guides, and all these influences reappear in the form taken by the theory that they conceive. Thus, it is not difficult to understand that a scientific theory can carry the marks of the time and place that saw its birth, that the work of Maxwell or of Thomson might be essentially English, and that it astonishes the French or the Germans.

What we have just said explains why the influence of a people to whom the author of a theory belongs, the context in which the author lives, and the era that sees the author's work make themselves felt much more in the erroneous or simply hypothetical parts of the theory than in the parts to which their logical form gives greater certainty.

In addition, an analogous remark may be made in every case in which one seeks to make precise the influence that a people or a context exerts on a human work. What in this work are subject to this influence are above all defects, those things through which the work participates in the prejudices and ignorance of the human community. On the other hand, whatever escapes this influence is what makes this work truly original and makes the author different from predecessors and contemporaries. It is what animates the breath of the mind, for, without regard for contexts or peoples, physical barriers or political frontiers, the mind breathes freely.

4

Some Reflections on the Subject of Experimental Physics[1]

Along with "The English School and Physical Theories," this article constitutes the core of Duhem's The Aim and Structure of Physical Theory. *It is the article that introduces the so-called Duhem thesis about the underdetermination of hypothesis by experiment: Physicists do not submit single hypotheses, but groups of hypotheses, to the control of experiment, and thus, experiment alone cannot conclusively falsify hypotheses. The various parts of this article are reworked and reordered in part II, chapters 4, 5, and 6 of the larger work.*

First Part
What Is an Experiment in Physics?

I. An experiment in physics is not simply the observation of a phenomenon; it is also the theoretical interpretation of that phenomenon.

What is an experiment in physics? Here is a question that no doubt will astonish more than one reader of the *Revue des questions scientifiques.* Is it necessary to ask the question and is the answer not obvious? To produce a physical phenomenon in conditions such that we can observe it exactly and minutely, with the aid of appropriate instruments, is that not the operation the whole world designates by the words "an experiment in physics"?

---

1. ["Quelques réflexions au sujet de la physique expérimentale," *Revue des questions scientifiques* 36 (1894): 179–229.]

Go into the laboratory. Approach a table cluttered with a mass of apparatus: an electric battery, copper wires wrapped in silk, beakers full of mercury, coils, a little iron bar carrying a mirror. An observer inserts the metallic stem of a rod with an ebonite[2] head into small holes. The iron vibrates and, through the mirror attached to it, reflects a luminous band on a celluloid scale, and the observer follows its movements. Here, without any doubt, is an experiment. The physicist minutely observes the vibrations of the piece of iron. Ask him what he is doing. Is he going to reply, "I'm studying the vibrations of the little iron bar with the mirror attached"? No. He will tell you that he is measuring the electrical resistance of a coil. If you are surprised, if you ask him what these words mean and what relation they have to the phenomena that he has observed, which you have observed at the same time, he will reply that your question would need too long an explanation, and you will be sent away to take a course on electricity.

In fact, the experiment that you have seen, like all experiments in physics, consists of two parts. It consists, in the first place, of an observation of certain phenomena. To make this observation, it is enough to be attentive and to have sufficiently quick senses. It is not necessary to understand physics. In addition, the experiment consists in the *interpretation* of the observed facts. To be able to make this interpretation, it is not sufficient to be alert and to have a practiced eye. You must know the accepted theories. You must understand how to apply them. You must be a physicist. Anyone who sees clearly can follow the movements of a spot of light on a transparent scale and see if it moves to the right or the left, or if it comes to rest at such and such a point. There is no need to be very learned to be able to do that. But if you are ignorant of electrodynamics, you will not be able to complete the experiment. You will not be able to measure the resistance of the coil.

Let us take another example. Regnault studies the compressibility of gases. He takes a certain quantity of gas, shuts it in a glass tube, keeps the temperature constant, and measures the pressure of the gas and the volume it occupies. Here, we might say, is the minute and precise observation of certain phenomena, or certain facts. Assuredly, in front of Regnault, between his hands, between the hands of his assistants, facts are produced. Is it the record of these facts that Regnault has left behind to

2. [Ebonite, or vulcanite, a hard, black substance made by heating India rubber and sulfur, was commonly used as an insulator in nineteenth-century electrical experiments.]

contribute to the advancement of physics? No. In a viewfinder, Regnault has seen the image of a certain surface of mercury come level with a certain mark. Is this what he has written down in the record of his experiments? No, he has written down that the gas occupies a volume having a particular value. An assistant has raised and lowered the lens of a cathetometer[3] until the image of another surface of mercury has just come level with the crosshairs of an eyepiece. He has then observed the arrangement of certain marks on the vernier of the instrument. Is this what we find in Regnault's papers? No, we read there that the pressure supported by the gas has a particular value. Another assistant has seen the mercury come level with a certain constant mark in a thermometer. Is this what has been recorded? No, they noted that the temperature was fixed and attained a particular magnitude. Now, what is the magnitude of the volume occupied by the gas? What is the value of the pressure that it supports? What is the magnitude of the temperature it reaches? Are these facts? No, these are three abstractions.

To form the first of these abstractions—the magnitude of the volume occupied by the gas—and to make it correspond to the observed fact, that is to say, the leveling of the mercury with a certain mark, it was necessary to measure volume of the tube. It was necessary to appeal not only to abstract notions of arithmetic and geometry, and the abstract principles these sciences rest on, but also to the abstract concept of mass and to hypotheses of general mechanics and celestial mechanics that justify the use of a balance in the comparison of masses. To form the second—the value of the pressure supported by the gas—it was necessary to use concepts that are exceedingly profound and difficult to acquire: pressure, cohesive force. It was necessary to call on the aid of the mathematical laws of hydrostatics, themselves founded on general mechanics. It was necessary to bring into play the law of compression for mercury, whose determination relies on the most delicate and controversial questions in the theory of elasticity. To form the third, it was necessary to define temperature and to justify the use of the thermometer. All those who have studied the principles of physics with some care know how distant the concept of temperature is from the facts and how difficult it is to capture.

Hence, when Regnault made an experiment, he had some facts before his eyes. He observed phenomena. But what he has transmitted to us about this experiment is not the recitation of the observed facts but some abstract

---

3. [An instrument for measuring small differences in height—for example, the difference between the surfaces of two mercury columns confined in glass tubes.]

data that accepted theories permitted him to substitute for the concrete evidence he had really gathered.

What Regnault did is what every experimental physicist necessarily does. That is why we are able to state the following principle, whose consequences will be developed in the present study: *An experiment in physics is a precise observation of a group of phenomena, accompanied by the* INTER-PRETATION *of these phenomena. This interpretation replaces the concrete data really gathered by observation with abstract and symbolic representations that correspond to them by virtue of physical theories accepted by the observer.*

II. This type of experiment characterizes sciences that have arrived at the phase called *rational*.

In declaring that the interpretation of facts by means of theories admitted by the observer forms an integral part of experimentation in physics and that it is impossible, in such an experiment, to dissociate or separate these statements of facts and the transformation that theory makes them undergo, perhaps we are going to scandalize more than one mind worried about the rigor of science. More than one person is going to raise against us rules sketched a hundred times by philosophers and experimenters from Bacon to Claude Bernard, from the *Novum Organum* to the *Introduction à la médicine expérimentale* (*Introduction to Experimental Medicine*). Let theory suggest experiments to perform; nothing could be better. Once the experimentation has been made and the results have been clearly observed, let theory take hold of them to generalize them, coordinate them, and draw from them new subjects of experiment; again, nothing could be better. But as long as the experiment lasts, theory ought to wait at the door of the laboratory, remaining silent and leaving the scientists face to face with the facts, without bothering them. These should be observed with no preconceived ideas, gathered with the same minute impartiality whether they confirm the predictions of the theory or contradict them. The record that the observer will give us of his experiment ought to be a faithful and scrupulously exact image of the phenomena. It should not even allow us to guess which system the scientist has confidence in or distrusts.

This rule is good for certain sciences—for those in which it is possible to apply it.

Here, for example, is a physiologist. He accepts that the anterior roots of the spinal cord enclose the motor nerves and the posterior roots enclose the sensory nerves. The theory he accepts leads him to imagine an experiment. If he cuts such an anterior root, it ought to suppress the corresponding part of the body's ability to move without removing its ability to feel sensations. When he observes the results and reports them, after having sectioned this root, he must set aside all his ideas concerning the physiology of the spine, his report must be a brutal tracing of facts. He is not permitted to silently pass over a movement, or quiver, contrary to his predictions. He is not permitted to attribute it to some secondary cause unless a special experiment has made this cause evident. He must, if he does not wish to be accused of scientific bad faith, establish an absolute separation, a water-tight compartment, between the consequences of his theoretical deductions and the results of these experiments.

This method is appropriate to sciences that are still close to their origins—such as physiology, such as certain branches of chemistry—and to sciences in which the researcher observes the facts directly, in which he reasons immediately about the observed facts. It is not applicable to more advanced sciences that have arrived at a state of development in which mathematical instruments play an essential role—physics, for example. Scientists at the beginning of the century labeled this phase, perhaps improperly, with the epithets *analytic* and *rational*.

Today, the number and complexity of experimental facts, and the multitude of laws that constitute physics, would form an inextricably chaotic mass, if the human mind had not found a means to unravel this enormous mass of evidence, to classify it and to translate it into clear and concise language. This means is provided by the application of physical theories. We have explained elsewhere[4] how these theories substitute a sort of symbolic representation, or schema, formed of elements borrowed from algebra and geometry for the properties of bodies whose variations constitute physical phenomena and for the experimental laws that govern these phenomena. Physical theories are the vocabulary that makes a magnitude correspond to every physical property and an equation correspond to every physical law.

At this point, the use of this vocabulary is indispensable to physicists, who would find it impossible to state the least law or to report the least observation without it. Take a typical experiment—for example, the experiment of Regnault that we spoke about a moment ago. Try to present it by ridding your language of all the abstract expressions introduced by

---

4. "Quelques réflexions au sujet des théories physiques," *Revue des questions scientifiques*, 2nd series, 1 (1892); "Physique et métaphysique," *Revue des questions scientifiques*, 2nd series, 2 (1893); "L'Ecole anglaise et les théories physique," *Revue des questions scientifiques*, 2nd series, 2 (1893). [These works are chapters 1, 2, and 3 of the present volume, respectively.]

physical theories, such as: pressure, temperature, density, optical axis of an eyepiece, coefficient of expansion, and so forth. You will realize that reporting this single experiment would require a volume so utterly confusing that it would unhinge the most attentive intellect. Or, rather, you will realize that this cannot be achieved. Just as French speakers, accustomed to their native language, cannot conceive a thought without at the same time stating it in French, so also physicists no longer conceive an experimental fact without simultaneously making it correspond to the abstract and schematic expression that theory gives it. This is why physicists say they are measuring the pressure of a gas when they see a black mark on a white background through a tube carrying rounded glass spheres. This is why they declare that they are determining the electrical resistance of a coil when they put copper rods in little holes and look at a line of light moving on a scale made of horn. It is a fantasy to pretend to separate the observation of physical phenomena from all theory or to boast of having written a *purely experimental* physics article. You might just as well attempt to state an idea without using any spoken or written sign.

Indeed, physicists are not the only people who must call on theories to state the results of their experiments. Chemists or physiologists, when they make use of the instruments of physics, such as the thermometer, the manometer, the calorimeter, or the galvanometer, implicitly admit the accuracy of the theories that justify the use of these instruments. These theories give meaning to the abstract concepts of pressure, temperature, quantity of heat, and intensity of current that replace the concrete readings of the instruments. But these theories they use, like the instruments they employ, are from the domain of physics. Chemists and physiologists place their confidence in physicists by accepting, along with the instruments, the theories without which their readings would be stripped of meaning. It is the physicist they suppose to be infallible. Physicists, on the other hand, are obliged to rely on their own theoretical ideas or those of their peers. From the point of view of logic, the difference is of little importance. For physiologists and for chemists, as for physicists, stating the outcome of an experiment implies, in general, an act of faith in the accuracy of a whole group of theories.

There is more. To the extent that a science progresses and distances itself from simple empirical knowledge, or the statement of the crudest laws, the role played by theory in the interpretation of the facts of experience continues to grow. When a science begins, when it is no different from common sense made more attentive, the record of experimental facts that it establishes is an exact image of the reality observed. Physiology, in many of its parts, offers us the image of a science in this state. But the more a science progresses, the more considerable becomes the depth of the theoretical considerations that separate the concrete fact really found by the observer and the abstract, symbolic translation given to it. Take, for example, chemistry in its actual state. Take, in particular, those of its branches that are most perfectly developed: the chemistry of carbon compounds and organic chemistry. What a difference there is between experimental fact and theoretical interpretation, the symbolic translation given to it by a chemist! Experiment teaches us that by substituting the acid group CO-OH for one H in benzene, we obtain benzoic acid. Measure the distance separating this statement from the actual concrete observations that it represents, and you will understand that the more a science progresses, the more the symbolic translation it substitutes for the experimental facts is abstract and distant from the facts.

III. That an experiment in physics can never condemn an isolated
    hypothesis, but only a whole theoretical group

The physicist who gives an account of an experiment implicitly recognizes the accuracy of a whole group (*ensemble*) of theories. Let us admit this principle and see what consequences we may deduce when we seek to appreciate the role and the logical import of an experiment in physics.

To avoid all confusion, we will distinguish two sorts of experiments: *applied* experiments and *testing* experiments.

You are confronted with a problem in physics to be solved practically. To produce such and such an effect, you wish to make use of knowledge acquired by physicists. You wish, for example, to light an incandescent electric lamp. The accepted theories show you the means to resolve the problem. But to make use of these means, you must acquire certain information. You must, I suppose, determine the electromotive force of the battery at your disposal. You measure this electromotive force. This is an *applied experiment*. It is not the goal of this experiment to determine whether the accepted theories are accurate. It proposes simply to make use of these theories. To perform the experiment, you make use of instruments that legitimate these same theories. There is nothing there to offend logic.

But applied experiments are not the only ones that physicists have to perform. Science is able to aid in practical matters by means of these alone, but applied experiments have nothing to do with science's origin and development. Besides applied experiments, there are *testing experiments*.

Physicists dispute some law or call into doubt some theoretical point. How do they justify their doubts? How will they demonstrate the inaccuracy of the law? From the proposition under indictment, they draw out the prediction of an experimental fact. They establish the conditions under which the fact should be produced. If the fact is not produced, the proposition will be irrevocably condemned.

F. E. Neumann assumed that, in a ray of polarized light, the vibration is parallel to the plane of polarization. Many physicists called this proposition into doubt. What did O. Wiener do to transform this doubt into a certainty and condemn Neumann's proposition? He deduced from this proposition the following consequence. If a beam of light reflected from a glass surface is made to interfere with an incident beam, polarized perpendicularly to the plane of incidence, it must produce fringes parallel to the reflecting surface. He set up the conditions in which these fringes ought to be produced and showed that the predicted fringes were not produced. From this observation he concluded that the proposition of F. E. Neumann was false—that in a ray of polarized light, the vibration is not parallel to the plane of polarization.

Such a mode of demonstration seems as convincing, as irrefutable, as the reduction to absurdity customary among mathematicians. Moreover, this demonstration is patterned after reduction to absurdity, with experimental contradiction playing in the one the role that logical contradiction plays in the other.

In reality, the demonstrative value of the experimental method is not quite so rigorous or absolute. The conditions in which it functions are much more complicated than we have supposed. Understanding its results is much more delicate and subject to caution.

Physicists propose to demonstrate the inaccuracy of a proposition. To deduce from this proposition the prediction of a phenomenon, to set up the experiment that must show that this phenomenon appears or does not appear, or to interpret the results of this experiment and report that the predicted phenomenon was not produced, they are not limited to making use of the disputed proposition. They employ at the same time a whole group of theories accepted by them as beyond dispute. The prediction of the phenomenon whose nonproduction will cut off the debate does not derive from the disputed proposition taken in isolation but from the disputed proposition joined to this whole group of theories. If the predicted phenomenon does not appear, it is not the disputed proposition alone that is shown to be wanting—it is the whole theoretical scaffolding used by the physicists. Experiment teaches us only that there is at least one error among all the propositions used to predict this phenomenon and to establish that it has not appeared. But experiment does not tell us where this error lies. Do physicists declare that this error is precisely contained in the proposition that they wish to refute and not elsewhere? To do so is to admit implicitly the accuracy of all the other propositions they have used. To the degree that this confidence is justified, so is their conclusion.

Let us take, for example, O. Wiener's experiment. To predict the formation of fringes in certain circumstances, in order to show that these fringes do not appear, Wiener did not merely make use of Neumann's celebrated proposition, which he wished to refute. He did not merely assume that, in polarized rays, the vibrations are parallel to the plane of polarization; in addition, he used the propositions, laws, and hypotheses that constitute optics as it is commonly accepted. He assumed that light consists of simple periodic vibrations, that at every point the mean kinetic energy of the vibratory movement measures the intensity of the light; and that the exposure of a photographic plate measures the different degrees of this intensity. It is by joining these different propositions and a few others that it would take too long to enumerate to Neumann's that he was able to formulate a prediction and to recognize that the experiment refuted this prediction. If, according to Wiener, the refutation addressed only Neumann's proposition, which alone must carry the responsibility of the error that this refutation has made apparent, it is because Wiener regards the other propositions he invoked as beyond doubt. But this confidence does not impose itself as a logical necessity. Nothing prevents us from regarding Neumann's proposition as accurate and making some other commonly accepted hypothesis of optics carry the experimental contradiction. As Poincaré has shown, we may quite well extract Neumann's hypothesis from the clutches of Wiener's experiment on condition that in exchange, we abandon the hypothesis that takes the mean kinetic energy of the vibratory movement as a measure of the intensity of the light. Without being contradicted by experiment, we may leave the vibration parallel to the plane of polarization, as long as we measure the intensity of the light through the mean potential energy of the medium deformed by the vibratory movement.

These principles are so important that it will not perhaps be inappropriate to apply them to a second example. Let us choose again an experiment regarded as one of the most decisive in optics.

It is known that Newton devised an emission theory of optical phenomena. He supposed light to be formed of very small projectiles thrown out with great speed by the sun and other light sources. These projectiles penetrate all bodies. On account of the different parts of bodies among which they move, the projectiles submit to attractive or repulsive actions. When the distance that separates the interacting particles is very small, these actions are very powerful, and they vanish when the masses between which they are acting are appreciably separated. These essential hypotheses, joined to several others that we do not mention, lead to the formulation of a complete theory of the reflection and refraction of light. In particular, they lead to this consequence: The refractive index of light passing from one medium to another is equal to the velocity of the light projectile in the medium it enters, divided by its velocity in the medium it leaves.

It is this consequence that Arago chose in order to show that the emission theory contradicts the facts. From that proposition, in fact, another follows: Light travels faster in water than in air. Through a procedure that Arago indicated and that Foucault made practical, awkward though it was, let us compare the speed of light in water and the speed of light in air. We will find that the former is slower than the latter. We might, therefore, conclude, with Foucault, that the emission system is incompatible with the reality of the facts.

I say the emission *system* and not the emission *hypothesis*. In fact, what the experiment shows to be tainted with error is the whole group of propositions accepted by Newton and, after him, Laplace and Biot. It is the entire theory from which the relation between the refractive index and the speed of light in different media is deduced. But in condemning the system as a whole, in declaring that it is tainted by error, the experiment does not tell us where that error lies. Is it in the fundamental hypothesis that light consists of projectiles thrown out with great speed by luminous bodies? Is it in some other assumption concerning the actions that the luminous corpuscles undergo on account of the media through which they move? About this we know nothing. It would be bold to believe, as Arago seems to have thought, that Foucault's experiment permanently condemns the emission hypothesis itself: the assimilation of a ray of light to a swarm of projectiles. Who knows whether one day we will see a new optics founded on this supposition arise?

In summary, physicists can never submit an isolated hypothesis to the control of experiment, but only a whole group of hypotheses. When an experiment is in disagreement with their predictions, it tells them that at least one of the hypotheses which constitute this group is erroneous and must be modified, but it does not tell them which one must be changed.

We are now a good distance from the mechanism of experimentation as it is freely imagined by people who are unfamiliar with its functioning. It is commonly thought that each of the hypotheses that physics uses may be taken in isolation, submitted to the control of experiment, and then, when varied and numerous proofs have established its validity, put in place in an almost definitive manner in the totality of science. In reality it is not so. Physics is not a machine that lets itself be taken apart. We cannot test each piece in isolation and wait, in order to adjust it, for its soundness to have been minutely regulated. Physical science is an organism that must be taken as a whole. One part of the organism cannot be made to function without its most distant parts entering into play, some more, others less, but all to some degree. If some trouble, some difficulty, is revealed in its functioning, the physicist will be obliged to discover which organ needs to be straightened out or modified without it being possible to isolate this organ and to examine it on its own. Presented with a watch that does not work, the watchmaker takes apart all the little wheels and examines them one by one until the one that is bent or broken is found. Presented with a sick person, the doctor cannot perform a dissection to establish a diagnosis. The doctor must decide the seat of the illness only by inspecting the effects produced on the whole body. The physicist charged with reforming a defective theory resembles the doctor, not the watchmaker.

IV. Crucial experiments[5] are impossible in physics.

Let us press this point further, for we are touching on one of the essential points of the experimental method employed in physics.

Reduction to absurdity, which seems to be merely a means of refutation, may become a method of demonstration. In order to demonstrate that a proposition is true, it is sufficient to drive those who support the contrary proposition back to an absurd consequence. We know the profit that Greek geometers drew from this mode of proof.

Those who assimilate experimental contradiction to reduction to absurdity think that they may follow a method in physics similar to the one Euclid used in geometry. Do you wish to obtain a certain and incontestable theoretical explanation for a group of phenomena? List all the hypotheses that may be made to account for this group of phenomena. Then, by experimental contradiction, eliminate all of them except one. This last one will cease to be a hypothesis and will become a certainty. Suppose, in particular, that we are confronted with only two hypotheses.

5. [Duhem's expression throughout is the Latin *experimentum crucis*, which is etymologically connected with *cross* or *crossroads*.]

Search for experimental conditions such that one of the hypotheses predicts the appearance of a phenomenon and the other the production of a completely different phenomenon. Bring about these conditions and observe what happens. To the extent that you observe the first of the predicted phenomena or the second, you condemn the second hypothesis or the first. The hypothesis not condemned will henceforth be indisputable. Debate will be cut off. A new truth will be acquired by physics. Such is a crucial experiment.

We are confronted with two hypotheses concerning the nature of light. For Newton, Laplace, and Biot, light consists of projectiles emitted with an extreme velocity. For Huygens, Young, and Fresnel, light consists of vibrations, whose waves are propagated in an elastic medium. These two hypotheses are the only possibilities entertained. Either a motion is carried away by the body that it moves or else it passes from one body to another. Follow the first hypothesis: It tells you that light moves more quickly in water than in air. Follow the second: It tells you that light moves more quickly in air than in water. Set up Foucault's apparatus. Set the revolving mirror in motion. Two spots of light will form: one white, the other greenish. Is the greenish band to the left of the white band? If so, light moves more quickly in water than in air and the hypothesis of waves is false. Is the greenish band to the right of the white band? If so, light moves less quickly in water than in air, and it is the emission hypothesis that is condemned. You compare the position of the two bands. You see the greenish band to the right of the white band. The debate is decided. Light is not a body: It is a vibratory motion with waves propagated in an elastic medium. The emission hypothesis has had its day. The wave hypothesis has ceased to be doubtful: It is a new article in the scientific credo.

What we have said in the preceding paragraph shows how far it would be mistaken to attribute to Foucault's experiment so simple a meaning and so decisive an import. Foucault's experiment does not cut between two hypotheses, the emission hypothesis and the wave hypothesis. It cuts between two theoretical groups or systems, each taken as a whole, between Newton's optics and Huygens's optics.

But let us accept for a moment that in each of the two systems, everything was compelled or necessary from logical necessity except one solitary hypothesis. Let us admit as a consequence that in condemning one of the two systems, the facts uniquely condemn the lone doubtful assumption that one system contains. Does it follow from this that, just as the reduction of a theorem to absurdity assures the truth of the contradictory theorem, we might find in crucial experiments an irrefutable means of transforming one of the two hypotheses under consideration into certain truth? Between two contradictory propositions of geometry, there is no room for a third. If the one is false, the other is necessarily true. Is it the same for two hypotheses in physics? Will we ever dare to affirm that no other hypothesis is imaginable? Light may be a swarm of projectiles; it may be a vibratory movement propagating waves in an elastic medium. Can it be nothing but the one or the other of these two things? Perhaps Arago thought this, but it would be difficult for us to share his conviction, since Maxwell has proposed to attribute light to periodic electrical currents transmitted within a dielectric medium.

The experimental method is not able to transform a physical hypothesis into an indisputable truth, for we are never certain that we have exhausted all the imaginable hypotheses concerning a group of phenomena. Crucial experiments are impossible. The truth of a physical theory is not decided by the toss of a coin.

V. Consequences of the preceding principles for teaching physics

In general, you may imagine that each hypothesis of physics can be separated from the group and submitted to the control of experiment in isolation. Naturally, from this erroneous principle, false consequences are deduced about the method according to which physics should be taught. People would like teachers to lay out all the hypotheses of physics in a definite order: to take the first, to present it, to expound its experimental verifications, and, when these verifications have been recognized as sufficient, to declare the hypothesis accepted. They would have the same operation repeated on the second hypothesis, on the third, and so on in order until all of physics would be built up. Physics would be taught in the same way that geometry is taught. Hypotheses would be deduced in the same way as theorems. The experimental proof of each assumption would take the place of the demonstration of each proposition. Nothing would be proposed that had not been sufficiently justified by the facts. Such is the ideal that many teachers present and which several perhaps believe they have attained.

This ideal is false. This manner of conceiving the teaching of physics follows from an erroneous conception of experimental science. If the interpretation of the least experiment in physics supposes the employment of a whole group of theories, if even the description of this experiment requires a mass of abstract and symbolic expressions for which theories alone fix the sense and the correspondence with the facts, it would be wise for physicists to undertake the development of a long chain of hypotheses and deductions before attempting any comparison between the theoretical edifice and concrete reality. Again, in describing the experiments which verify the theories already developed, they would often be obliged to anticipate the theories to come. They would not be able, for example, to attempt the slightest experimental verification of the principles of dynamics before having not only developed the connections among the propositions of general mechanics but also having laid the foundations of celestial mechanics. Again, in reporting the observations that verify this group of theories, they must suppose the laws of optics to be known, since these alone justify the use of astronomical instruments.

Let teachers therefore develop, in the first place, the essential theories of science. No doubt, in presenting the hypotheses on which these theories rest, it is good that they indicate what is given from common sense (*sens commun*), the facts gathered by ordinary observation, which have led to the formulation of these hypotheses. But let them proclaim loudly that these facts, although sufficient to suggest hypotheses, are not sufficient to verify them. It is only after having constituted an extended body of doctrine— that is, after having constituted a complete theory—that they will be able to compare the consequences of this theory with experiment.

Teaching must make the student take hold of this primary truth: Experimental verifications are not the foundation of theory, they are its crown. Physics does not make progress in the way geometry does. The latter grows by means of the continual addition of new theorems, demonstrated once and for all, which are joined to the theorems that have already been demonstrated. The former is a symbolic picture, that is continually revised to give greater and greater extension and unity. The *whole* of it forms a more and more precise image of the *whole* of the experimental facts. At the same time, each detail of this image, cut off and isolated from everything else, loses all significance and no longer represents anything.

VI. That the result of an experiment in physics is an abstract and symbolic judgment

Beyond the reporting of a phenomenon or of a group of phenomena, all experiments in physics consist essentially of an interpretation which brings into play an entire group of theories admitted by the observer. This interpretation has the goal of replacing the concrete facts actually observed with abstract and symbolic representations. We have already deduced some consequences from the first portion of this principle, now let us examine the second part.

That the result of the operations to which experimenters devote themselves is not a fact but an abstract symbol is obvious to anyone who reflects. Open any article whatsoever on experimental physics and read its conclusions. These conclusions are not solely the repetition of certain facts. They are abstract statements to which you would not be able to attach any sense if you did not know the physical theories accepted by the author. You read there, for example, that the electromotive force of a particular battery increases by so many volts when the pressure sustained by the battery increases by so many kilograms per square centimeter. What does this statement mean? Those who are ignorant of physics and for whom this statement seems to be a dead letter may try to see here a simple manner of expressing a fact reported by the observer; in technical language incomprehensible to noninitiates but clear to initiates. This would be a mistake. It is true that initiates, those possessing the theories of physics, may be able to translate this statement into facts and to conduct the experiment with the corresponding outcome. But it is a remarkable thing that they are able to bring it about in an infinity of different ways. They may exert pressure by turning over mercury in a glass tube or by bringing into action a hydraulic press. They may measure this pressure with a manometer of free air, a manometer of compressed air, or a metal manometer. To understand the variation in the electromotive force, they might employ in turn all types of electrometers, galvanometers, and electrodynamometers. Each new arrangement of apparatus furnishes them with new facts to report. They might employ arrangements of apparatus that the author of the paper would not have suspected and see phenomena that the author would never have seen. Yet all these manipulations, so different that a noninitiate would not see any similarity between them, are not different experiments. They are only different forms of the same experiment. The facts produced are as dissimilar as possible. Yet the reporting of these facts is expressed by this single statement: The electromotive force of a particular battery increases by so many volts when the pressure increases by so many kilograms per square centimeter.

This statement, we see, is not the repetition of certain observed facts made in an abbreviated and technical language. It is the transposition of these facts into the abstract and schematic world created by physical theories. In this world the instrument before me is no longer an assembly of pieces of threaded copper, metal wires wrapped in silk and wound around a frame, and a little piece of steel suspended on a silk thread, but a *tangent galvanometer*; that is to say, the circumference of a circle carrying a current, with a magnetic needle at the center. In this world a battery is no longer a vase of pottery or glass, filled with certain liquids, in which certain solids are immersed, but a conceptual artifact (*être de raison*) symbolized by certain chemical formulae, a certain electromotive force, and a certain resistance.

VII. On approximation in physical experiments

Between an abstract symbol and a fact there may be correspondence, but there cannot be complete parity. The abstract symbol cannot be an adequate representation of the concrete fact; the concrete fact cannot be the realization of an abstract symbol. The abstract schema through which physicists express concrete facts recorded in the course of an experiment cannot be the exact equivalent or faithful retelling of their observations. It is a consequence of this, as we have seen, that very different concrete facts may build upon one another when they are interpreted by a theory and no longer constitute a single experiment expressed by a unique symbolic statement. Conversely, it also follows from this, as we will see, that a single group of concrete facts can in general be made to correspond not just to a single symbolic judgment but to an infinity of judgments differing from one another and logically incompatible among themselves.

In order to report the phenomena that appear in an experiment in physics, we have no other means than to fall back on the evidence of our senses: sight, hearing, or touch. However complicated, however perfect may be the instruments employed, their use leads back, in the last analysis, to the reports of our senses. But our senses have a limited sensitivity, and this is a truth of common sense. What falls below a certain lower limit escapes them. Ordinary language, shaped to fit what is given by the senses, lends to words a certain vagueness which translates the uncertainties of our perceptions.

It is not at all the same with a symbolic language created by physical theories. Thanks to the use of mathematical notions, this language expresses itself in judgments amenable to limitless rigor and precision. Also, there may be no exact equivalence between a fact reported by the senses, with the vagueness that a corresponding report carries, and a theoretical judgment stated in a mathematical form that excludes all ambiguity. In order to translate the uncertainty carried by the limited sensitivity of our perceptions into its own language, theory replaces the recitation of a group of facts not by a unique abstract judgment but by an infinity of judgments among which it leaves us free to choose, or, rather, among which we must not choose but must accept the whole group. These judgments are different from one another and irreconcilable among themselves. From the point of view of mathematical logic, one of them cannot be true unless the others are false. But take one or another of these theoretical propositions. Apply the accepted theories to it and deduce consequences that the instruments used in physics allow you to translate into perceptible facts. The senses would not be able to distinguish between the consequences deduced from the one and the consequences deduced from the other. This is why, whereas mathematics regards the two propositions as excluding each other, physics regards them as identical.

This essential truth for understanding the experimental method—the correspondence of one group of facts to an infinity of different theoretical propositions—is expressed in the following proposition: The results of an experiment in physics are only *approximate*. Fixing the approximation that the experiment brings with it marks the indeterminacy of the abstract and symbolic proposition through which physicists replace the concrete facts that they have really observed. It defines the bounds that this indeterminacy may not cross.

Let us clarify these general principles by giving an example.

An experimenter has made certain observations. He has translated them into this statement: An increase in pressure of 100 atmospheres makes the electromotive force of a gas battery increase by 0.0845 volts. It might be equally legitimate to say that it had made this electromotive force increase by 0.0844 volts or, again, that it had made it increase by 0.0846 volts. How can these different propositions be equivalent for physicists? For if a number is 845, it cannot at the same time be 844, any more than it can be 846. This is what physicists mean in declaring these three judgments identical to their eyes: Taking the value 0.0845 volts for the change[6] in the electromotive force as a point of departure, if they calculate the deviation of the needle of their galvanometer by means of accepted theories—that is to say, the sole fact that the senses are able to observe—they will find a certain value for this deviation. If they repeat the same calculation, taking as their starting point the value 0.0846 volts or the value 0.0844 volts for the change in the electromotive force, they will find other values for the deviation of the pointer. But the three deviations calculated in this way differ too little to be discriminable by vision. This is why physicists will not distinguish the three values of the increase of the electromotive

6. [Curiously, given the surrounding statements, Duhem here writes of "la diminution de la force électromotrice." We have replaced "diminution" with "change."]

force, 0.0844, 0.0845, and 0.0846 volts, whereas mathematicians regard them as incompatible among themselves.

Let us suppose that, by means of calculations based on accepted theories, all the values of the change in the electromotive force lying between 0.0840 volts and 0.0850 volts lead to consequences that the readings on the instrument used by the physicists cannot distinguish. Physicists would not be able to say that this increase is equal to 0.0845 volts, but only that it is one of the numbers lying between 0.0840 and 0.0850 volts. Or, rather, this increase might be equally well represented by any one of these numbers. They present this infinity of possible values to us all at once by writing, for example, that 100 atmospheres make the electromotive force of the battery increase by 0.0845 ± 0.0005 volts.

The degree of approximation of an experiment depends on two essential elements: the nature and accuracy of the instrument employed and the theoretical interpretation of the experiment.

That the degree of approximation of an experiment depends on the instrument used in mounting the experiment is shown clearly by the preceding explanation. Here are two distinct abstract judgments. Let us ask of the accepted theories what consequences follow from these two judgments when they are applied to a first instrument and what consequences follow when they are applied to a second instrument. These two judgments may be translated in one of the instruments by means of two different facts, but the senses may not be able to distinguish the one from the other; and in the other instrument by means of facts which the senses distinguish without difficulty. Although they are equivalent for physicists who make use of the first apparatus, these two judgments will no longer be equivalent for physicists who makes use of the second. This truth is too clear for it to be necessary to insist upon it.

But perfecting an instrument is not the only element that increases the precision of an experiment. We may also increase this precision by perfecting the theoretical interpretation, that is, by eliminating *causes of error* by making appropriate *corrections*—and this is what remains to be explained.

VIII. On corrections and causes of error in physical experiments

To the extent that physics progresses, we see the decreasing indeterminacy of the group of abstract judgments that the physicist correlates with concrete facts. The accuracy of experimental results continues to grow, not only because those working on them furnish more and more precise instruments but also because physical theories yield more and more satisfactory rules for establishing the correspondence between facts and the schematic ideas that serve to represent them. This increasing precision is achieved, it is true, through increasing complication, through the obligation to observe a series of additional facts at the same time that the principal fact is being observed, and through the necessity of submitting the raw data of experience to manipulations and transformations that are more and more numerous and delicate. These transformations to which we submit the immediate data of experience are *corrections*.

If physical experiments were simple reports of fact, it would be absurd to add corrections to them. When the observer had looked carefully and minutely, nothing would remain but to say, "Here is what I saw." It would be inappropriate to reply: "You have seen something, but it is not what you were supposed to see. Allow me to make various calculations that will tell you what you ought to have seen."

On the other hand, the logical role of corrections may be understood quite well when we remember that a physical experiment is the reporting of a group of facts, followed by the translation of these facts into a symbolic judgment by means of rules borrowed from physical theories.

Before the physicist is an instrument, a set of concrete bodies. This is the instrument that the physicist manipulates. It is on this instrument that sensory reports are made—that is, *readings*, the basis of experiment. In order to interpret the experiment, the physicist does not think about this instrument but about a schematic instrument, which is not an assembly of concrete bodies at all, but a group of mathematical concepts of perfect solids or perfect fluids having a certain density, a certain temperature, and subjected at every point to a certain force represented by a geometrical magnitude.

This schematic instrument is not and cannot be the exact equivalent of the real instrument. But we understand that it might be possible to give a more or less accurate image of it. We understand that, after having thought about a schematic instrument that is too simple and too far from reality, the physicist seeks to substitute for it a schema that is more complicated but more representative. This passage from one schematic instrument to another which better symbolizes the concrete instrument is essentially the operation that the word *correction* designates in physics.

An assistant to Regnault gives him the height of a column of mercury contained in a manometer. Regnault corrects it. Does he suspect that his aide has not looked properly, that he has made a mistake? No. He has full confidence in the readings that have been made. If he did not have this confidence, he could not correct the experiment. He could do nothing but repeat it. So if, for this height read by his aide, Regnault substitutes another, it is by virtue of reasoning intended to diminish the difference between the abstract manometer, an ideal that exists only in his mind and to which his calculations are referred, and the concrete manometer, made of glass and mercury, which is before his eyes and on which his assistant makes readings. Regnault might be able to represent this real manometer through an ideal manometer made of incompressible mercury, which has the same temperature everywhere, and is subject at every point on its free surface to an atmospheric pressure that is independent of altitude. The difference between this overly simple schema and the reality would be too great, however, and, in addition, the experiment would be insufficiently precise. Hence, he conceives a new ideal manometer, more complicated than the first but representing the real and concrete manometer better. He supposes the new manometer to be formed of a compressible fluid; he supposes that the temperature varies from one point to another; and he admits that the barometric pressure changes when we go higher in the atmosphere. All these retouchings of the primitive schema constitute so many corrections: a correction for the compressibility of mercury, a correction for the unequal heating of the mercury column, and Laplace's correction for the barometric height. All these corrections have the effect of increasing the precision of the experiment. The physicist who complicates the theoretical representations of the observed facts with these corrections to enable that representation to come closer to reality is similar to an artist who, after having made a sketch, adds shadows to show the model's relief better on a flat surface.

Those who see nothing but reports of facts in physical experiments do not understand the role played by corrections in these experiments. Furthermore, they do not understand what is meant in speaking of the *causes of error* that make up an experiment.

To allow a cause of error to remain in an experiment is to omit a possible correction which would increase the precision of the experiment. It is to be content with an overly simple theoretical representation when we could replace it with a more complicated but more perfect image of reality. It is to be content with a sketch when we could make a complete picture.

In his experiments on the compressibility of gas, Regnault allowed a cause of error to remain that he did not recognize but that has since been noted. He neglected the effect of its own weight on the gas under pressure. What might we mean by reproaching Regnault for not having taken this effect into account or for having omitted this correction? Might we mean that his senses had deceived him while he was observing the phenomena produced in front of him? Not at all. He is criticized for having overly simplified the theoretical image of the facts in representing the gas under pressure as a homogenous fluid, when by regarding it as a fluid with a density that varies with the height according to a certain law, he would have obtained a new abstract image, more complicated than the first, but one that would have reproduced reality better.

IX. On criticizing a physical experiment; how it differs from examining ordinary testimony

An experiment in physics being something completely different from the simple reporting of a fact, we understand without difficulty that the truth or certainty of an experimental result must be something of a completely different order from the truth or certainty of a reported fact. These types of certainty are so different in nature that they are evaluated by means of entirely different methods.

When sincere witnesses affirm the report of a fact, if their minds are sound enough not to confuse the play of the imagination with perceptions, and if they understand the language they use well enough to express their thoughts clearly, then the fact is certain. If I swear to you that on such and such a day, at such and such an hour, in such and such a street, I saw a white horse, unless you have reason to consider me a liar or subject to hallucinations, you ought to believe that on that day, at that hour, and in that street, there was a white horse.

The confidence which ought to be accorded to propositions stated by a physicist as the result of an experiment is not of the same nature. If physicists limit themselves to telling us the facts that they have seen, in the strict sense of seeing with their own eyes, their testimony should be examined according to the general rules appropriate for determining the degree of confidence merited by the testimony of individuals. If they are recognized to be in good faith—and this is, I think, the general case—their testimony should be accepted as an expression of the truth.

But to repeat, what physicists state as the result of an experiment is not the recitation of reported facts. It is the interpretation of these facts and their transposition into the abstract symbolic world created by theories that physicists regard as established.

Thus, after having submitted the testimony of a physicist to the rules that establish the degree of confidence merited by the recitation of a piece of testimony, you would have completed only one part—the easiest part— of the critical appraisal of an experiment.

In the first place, ask yourself with great care which theories the physicists regard as established and which they have employed in the interpretation of the facts they have reported. Unless you know these theories, it
will be impossible for you to grasp the meaning they attribute to their
statements. The physicists would stand before you like witnesses before a
judge who did not speak the same language.

If the theories accepted by a physicist are the same as those you accept,
and if you both agree to follow the same rules in the interpretation of the
same phenomena, then you speak the same language and you will be able
to understand each other. But this is not always the case. It is not so when
you discuss the experiments of a physicist who does not belong to your
school. It is not the same, above all, when you discuss the experiments of
a physicist who is separated from you by fifty years, a century, or two centuries. Then you must seek to establish a correspondence between the theoretical ideas of the author you are studying and your own and to interpret
anew what was interpreted by means of symbols that were accepted then,
by means of symbols that you accept now. If you arrive at this point, the
discussion of that author's experiment becomes possible for you. This
experiment will be given in a language alien to your own but one whose
vocabulary you possess. You will be able to examine it.

Newton, for example, made certain observations concerning the phenomena of colored rings. He interpreted these observations by means of
the optical theory that he had created, the emission theory. He interpreted
the rings as giving the distance between the "fit" of reflection and the "fit"
of easy transmission for light corpuscles of each color.[7] When Young and
Fresnel revived the wave theory in order to replace the emission theory, it
was possible for them to make elements of the new theory correspond to
elements of the old theory at certain points. They established, in particular, that the distance between a "fit" of easy reflection and a "fit" of easy
transmission corresponded to one-quarter of what the new theory called a
wavelength. Thanks to this remark, the results of Newton's experiments
could be translated into the language of waves. The numbers that Newton
had given, multiplied by four, gave the wavelengths of the various colors.

On the other hand, if you are unable to obtain sufficient information
about the theoretical ideas of the physicists whose experiments you are
discussing, if you are unable to establish a correspondence between the
symbols that they adopted and the symbols furnished by the theories you
accept, if you are unable to translate into your language the propositions

7. [For an explanation, see chapter 8, section XXIV.]

representing the results of their experiments, these results will be neither true nor false for you. They will be devoid of meaning, a dead letter. How many observations accumulated by physicists of former ages are thus lost forever! Their authors neglected to inform us about the methods they used to interpret the facts. It is impossible for us to transpose their interpretations into our own theories. They have sealed in their ideas in signs for which we lack the keys.

Perhaps these first rules seem naive, and you may be surprised to see us insisting on their employment. If these rules are commonplace, however, their lack is still more commonplace. How many scientific discussions there are in which each of the two sides claims to destroy its adversary through the irrefutable testimony of facts and in which contradictory observations are opposed to one another: The contradiction never exists in reality, which is always self-consistent. The contradiction is between the theories that each of the two champions uses to express this reality. How many propositions are designated monstrous errors in the writings of those who preceded us! Perhaps we might celebrate them as great truths if we would only inform ourselves about the theories that gave these propositions meaning and take care to translate them into the language of the theories advocated today.

But let us suppose that you have noted the agreement between the theories accepted by an experimenter and those that you regard as accurate. It would have to be the case that you could immediately make your own the judgments through which the results of the experiments were stated. You must now determine whether the rules imposed by the theories that you both accept in the interpretation of the observed facts were correctly applied by the experimenter; that is, whether all the necessary corrections were made. Often you will find that the experimenter has not taken into account all the legitimate possibilities. In applying the theories, an error in reasoning or in calculation may have been committed. An indispensable correction may have been omitted and a cause of error which ought to have been eliminated may have been allowed to remain.

The experimenter has employed the theories that you both accept in order to interpret the observations. The rules prescribed by these theories in this interpretation have been correctly applied by the experimenter and the causes of error have been eliminated or their effects corrected. This is still not enough to enable you to accept these experimental results. As we have said, the abstract propositions that the theories make correspond to the observed facts are not entirely determinate. An infinity of different propositions may correspond to the same facts; an infinity of evaluations

may express themselves through different numbers. The degree of indeterminacy possible in the abstract mathematical proposition expressing
the result of an experiment is precisely the degree of approximation of the
experiment. You need to know the degree of approximation of the experiment that you are examining. If the experimenter has indicated it, you
need to inform yourself about the reasoning used to evaluate it. If the
experimenter has not indicated it, you must determine it through your
own analysis. This estimation of the degree of approximation that makes
up a given experiment is a delicate operation. It is often so complicated
that an entirely logical order is difficult to maintain. The reasoning must
make room for that rare subtle quality, that sort of flair which is called
experimental sense—an endowment of the subtle mind (*esprit de finesse*)
more than of the geometrical mind (*esprit géométrique*).

The mere description of the rules governing the examination of an
experiment in physics and its acceptance or its rejection suffice to make
this essential truth apparent: The result of an experiment in physics does
not possess the same order of certainty as a fact reported by nonscientific
methods, through the mere sight or touch of a person healthy in body and
mind. The certainty of an experiment in physics is less immediate. It is
submitted to disputes that ordinary testimony escapes, and it always
depends on our confidence in a whole group of theories.

X. Less certain than the nonscientific statement of a fact, a physical
   experiment is more precise.

Although the report of a physical experiment lacks the immediate, and rel-
atively easy to establish, certainty of ordinary, nonscientific testimony, it
improves on the latter by virtue of the number and the minute precision of
the details that it makes known to us.

Ordinary testimony, the reporting of a fact established through commonsense procedures and not through scientific methods, can scarcely be
certain except on condition of not being detailed, or not being painstaking,
or only describing the fact roughly, in terms of what stands out most about
it. In such and such a street of the town, at such and such an hour, I saw a
white horse. This is all I might be able to assert with certainty. Perhaps, to
this general assertion, I might be able to add some particularity which
attracted my attention to the exclusion of other details: something strange
in the posture of the horse, or an oddity about its harness. But pressing me
with questions will be of little use. My memories would be confused and
my answers vague. Soon enough I would be reduced to telling you: "I do

not know." With few exceptions, common testimony has all the more certainty the less precise it is, the less it analyzes, and the more it confines itself to the roughest and most obvious considerations.

The record of a scientific experiment is quite different. It is not content with making a phenomenon known to us roughly. It claims to analyze it, making the smallest detail and the most painstaking particular known to us, and marking the rank and relative importance of each detail and peculiarity. This claim would exceed its powers, as they exceed the powers of ordinary observation, if the one were not better equipped than the other. The number and the minuteness of the details that accompany and compose each phenomenon would baffle the imagination, exceed the memory, and defy the power of speech if physicists did not have mathematical theory at their disposal. Mathematical theory is a marvelous instrument of classification and expression, and an admirably clear and compact symbolic representation. Physicists also have at their disposal numerical evaluation, or measurement, which provides them with an exact and brief means of estimation with which to indicate the relative importance of each particular. As we have already said, if, for a wager, someone undertook to describe a real experiment in physics while excluding all theoretical language, that person would fill an entire volume with the most tangled, confused, and incomprehensible account imaginable.

Therefore, if theoretical interpretation removes the immediate, indisputable certainty offered by the data of ordinary observation from the results of experiment in physics, in return, theoretical interpretation allows scientific experiment to penetrate further than common sense in the analysis and detailed description of phenomena.

Second Part
What Is a Law of Physics?

I. The laws of physics are symbolic relations.

Just as commonsense laws (*lois de sens commun*) are founded on the observation of facts through means natural to humans, physical laws are founded on the results of experiments in physics. It goes without saying that the profound differences that separate a fact established by nonscientific means from a fact that results from physical experiment also separate commonsense laws from physical laws. Thus, almost everything we have said about experiments in physics might be extended to the laws that science states.

Let us take a commonsense law, one of the simplest, as it is one of the most certain: All humans are mortal. This law, assuredly, relates abstract terms among themselves: the abstract idea of humanity in general and not the concrete idea of such and such a particular human being; the abstract idea of death and not the concrete idea of such and such a form of death. Only on condition that it relates abstract terms can the law be general.

But these abstractions are by no means symbols. They simply extract what is general in the concrete realities to which the law applies. Thus, in each of the particular cases in which we apply the law, we find concrete objects that realize these abstract ideas. Each time we wish to apply the law "All humans are mortal," we find ourselves in the presence of a particular individual instantiating the idea of humanity in general and a certain particular death instantiating the idea of death in general.

Matters are not the same for physical laws. Let us take one of these laws, Mariotte's law,[8] and examine its formulation without concerning ourselves, for the moment, with the accuracy of this law. At a constant temperature, the volume occupied by a given mass of gas is inversely related to the pressure it is subject to—such is the statement of Mariotte's law. The terms it introduces, the ideas of mass, temperature, and pressure, are still abstract ideas. But these ideas are not only abstract, they are also symbolic. Place yourself in front of a real, concrete case to which you wish to apply Mariotte's law. You are not dealing with a certain definite temperature instantiating the idea of temperature in general but with some warmer or cooler gas. You would not have a certain particular pressure instantiating the notion of pressure in general before you but some mercury in a tube of glass. No doubt a certain temperature corresponds to this warmer or cooler gas, and a certain pressure corresponds to this mercury in a tube of glass, but this correspondence is of a thing signified by the sign that replaces it or of a reality to the symbol representing it.

Because the abstract terms of a commonsense law are no more than what there is of a general nature in concrete, observed objects, the passage from concrete to abstract is made through an operation so necessary and so spontaneous that it remains unconscious. Placed in the presence of a certain man and a certain case of death, I connect them immediately to the

8. [In English-speaking countries, the following principle is commonly referred to as Boyle's law. Edme Mariotte (c. 1620–1684) was a founding member of the Academie des Sciences in Paris. In *De la nature de l'air* (Paris, 1679), he noted that the volume of a gas varies inversely with its pressure. Boyle's formulation of the same law appeared perhaps in 1660.]

general idea of humanity and to the general idea of death. This sudden, unconscious operation yields unanalyzed general ideas, abstractions taken, so to speak, *en bloc*. Without doubt, thinkers can analyze these general, abstract ideas. They can seek to deeply penetrate the meaning of the word *humanity*, and the meaning of the word *death*. This effort will lead to a better grasp of the law's metaphysical foundation (*raison d'être*). But this effort is not necessary to understand the law. For that, it is sufficient to take the terms which it relates in their obvious sense. Thus, this law is clear for all whether or not they are philosophers.

The symbolic terms connected by a law of physics are no longer abstractions that gush spontaneously from concrete reality. They are abstractions produced by a slow, complicated, conscientious work of analysis, the age-old work that has elaborated physical theories. If you have not undertaken this work, if you do not understand physical theories, you cannot understand a law or apply it. Depending on whether you adopt one theory or another, a law changes its meaning in such a way that it may be accepted by one physicist who admits a certain theory and rejected by another who accepts another theory. Take a peasant who has never analyzed the concept of humanity or the concept of death and a metaphysician who has spent his life analyzing them. Take two philosophers who have analyzed them and who have adopted different, irreconcilable definitions. For all these people, the law "All humans are mortal" will be equally clear and equally true. On the other hand, take two physicists who do not define pressure in the same manner because they do not admit the same theories of mechanics. One, for example, accepts the ideas of Lagrange; the other adopts the ideas of Laplace and Poisson. Submit to these two physicists a law whose statement brings into play the notion of pressure. They will hear the statement in two different ways. To compare it with reality, they will make different calculations so that one will find this law verified by facts which, for the other, will contradict it. Here is evident proof of the following truism: A law of physics is a symbolic relation whose application to concrete reality demands that we understand it and that we accept a whole group of theories with it.

II. That a physical law is, properly speaking, neither true nor false but approximate

A commonsense law is a simple, general judgment. This judgment is true or false. For example, take this law of common experience: In Paris, the sun rises each day in the east, climbs into the sky, then descends and sets in the west. Here is a true law without conditions or restrictions. On the other hand, let us take this statement: The moon is always full. That is a false law. If the truth of a commonsense law is placed in question, we can respond to this question with a yes or a no.

It is not the same with the laws that a fully mature physical science states in the form of mathematical propositions. Such laws are always symbolic. But a symbol is, properly speaking, neither true nor false. It is better or less well chosen to signify the reality that it represents. It delineates reality in a manner more or less precise and more or less detailed. But applied to a symbol, the words *truth* and *error* no longer have meaning. Thus, to those who ask if such a law of physics is true or false, the logician concerned about the strict meanings of the words will be obliged to respond, "I do not understand your question." Let us comment on this reply, which may seem paradoxical but the understanding of which is necessary to those who aspire to know what physics is.

As practiced by physics, the experimental method does not make a single symbolic judgment correspond to a given fact but to an infinity of different symbolic judgments. The degree of indeterminacy of the symbol is the degree of approximation of the experiment in question. Take a group of analogous facts. For the physicist to find a law for these facts will be to find a formula that contains the symbolic representation of each of these facts. The indeterminacy of the symbol that corresponds to each fact carries with it, from then on, the indeterminacy of the formula which must bring together all these symbols. We can make an infinity of different formulas and an infinity of distinct physical laws correspond to the same collection of facts. To be accepted, each of these laws must make each fact correspond not to the symbol for that fact but to some one among the infinite number of symbols that are capable of representing the fact. This is what is meant when the laws of physics are said to be nothing but approximations.

Let us imagine, for example, that you might not be content with the information supplied by this commonsense law: In Paris, the sun rises each day in the east, climbs into the sky, then descends and sets in the west. You turn to the physical sciences for a precise law of the motion of the sun seen from Paris, a law indicating to the Parisian observer what position the sun occupies in the sky at each moment. In order to resolve this problem, the physical sciences will not use realities accessible to the senses (the sun such as you see it shining in the sky), but symbols through which theories represent these realities. Despite the sun's surface irregularities and the enormous prominences that it carries, those sciences will replace the real sun with a geometrically perfect sphere, and it is the position of the center of this ideal sphere which they will seek to try to determine. Or, rather, they will seek to determine the position that this point would occupy if astronomical refraction did not deflect the rays of the sun and if annual aberration did not modify the apparent position of the stars. It is therefore only a symbol that physical sciences substitute for the single reality accessible to the senses offered for your observations, to the shining disk that your telescope can view. To make the symbol and the reality correspond, it is necessary to take complicated measurements. You must make the edge of the solar image coincide with the crosshair of a micrometer. You must take multiple readings on divided circles. You also need calculations whose legitimacy depends on the theories you accept; on the theory of atmospheric refraction and the theory of aberration.

Your formulas will not yet grasp the point symbolically named the center of the sun. What they will grasp are the coordinates of this point, its right ascension and declination. And the meaning of these coordinates cannot be understood unless the laws of cosmography are understood.

But, supposing that the corrections for aberration and refraction have been made, can we make a single value for the right ascension and a single value for the declination of the center of the sun correspond to a definite position of the solar disk? Not at all. The optical power of the instrument that you use to see the sun is limited. The different observations that make up your experiment, and the different readings they require, are limited in accuracy. You would not be able to tell the difference between the solar disk in one position versus another if the variation were small enough. Let us posit that you will not be able to distinguish two points when the angular distance separating them is less than one second. It will be sufficient for you to determine the position of the sun at a given moment to know the right ascension and declination of the center of the sun to within one minute. From then on, to represent the movement of the sun, which at each instant occupies only a single position, you would be able to give not one value for the right ascension and declination but an infinity of values for the right ascension and an infinity of values for the declination. At any moment, only two acceptable values for the right ascension or two acceptable values for the declination may not differ by more than one second.

Now look for the law of motion of the sun—that is, two formulas that permit you to calculate, for every instant during the day, the value of the right ascension of the sun's center and the value of the declination of the same point. Is it not obvious that, to represent the change in the right ascension as a function of time, you could adopt not one unique formula but an infinity of different formulas, as long as, at any instant, all these formulae lead you to values of the right ascension differing among themselves by less than one second? Is it not evident that things will be the same for the declination? Hence you might equally well represent your observations on the motion of the sun by an infinity of different laws. These various laws are expressed through equations that analysis regards as incompatible. If any one among these equations is verified, no other can be. Yet, for the physicist, all these laws are equally acceptable, for they determine all the positions of the sun with an accuracy greater than that available from observation. The physicist has no right to say that any of these laws is true to the exclusion of the others.

No doubt, physicists have the right to choose among these laws, and in general they will choose between them. But the motives that guide their choice will have neither the same nature nor the same imperious necessity as those which require the preference of truth over error. They will choose a certain formula because it is simpler than the others. The feebleness of our minds constrains us to attach great importance to considerations of this kind. But we are no longer in an era that assumes that the intelligence of the Creator is constrained by the same disability, or one that rejects any law expressed in an algebraic equation that is too complicated, in the name of the simplicity of the laws of nature. Physicists especially will prefer one law to another when the first follows from theories that they accept. For example, they will ask the theory of universal attraction to decide which formulas they should prefer among all those able to represent the motion of the sun. But physical theories are only one means of classifying and relating the approximate laws to which experiments are subject. Theories cannot, therefore, modify the nature of one of these laws and confer absolute truth on it.

Thus, any law of physics is an approximate law. As a consequence, for the strict logician, it can be neither true nor false. Any other law that represents the same experiments with the same approximation may aspire, as legitimately as the first, to the title of the true law or, to speak more rigorously, the acceptable law.

III. That all laws of physics are provisional

The essential character of a law is its permanence. A proposition is not a law unless it is true today and will still be true tomorrow. Would it not be a contradiction to say that a law is provisional? Yes, if we mean by *law* those laws that common sense reveals to us and that might, in the strict meaning of the word, be called true. Such laws cannot be true today and false tomorrow. No, if we mean by *law* the laws that physics states in a mathematical form. Such laws are always provisional. We do not mean that laws in physics are true for a certain time and false later, for at any moment they are neither true nor false. Such laws are provisional because they represent the facts to which they apply with an approximation that physicists judge sufficient but which will one day cease to satisfy them.

As we have already noted, the degree of approximation of an experiment is not something fixed. It increases to the extent that instruments become more accurate, that the causes of error are more strictly avoided, or that more precise corrections permit their better evaluation. To the extent that experimental methods improve, the indeterminacy of abstract symbol, brought into correspondence with a concrete fact by an experiment in physics, will diminish. Many symbolic judgments which have been regarded at one time as a good representation of a definite concrete fact will no longer be accepted in another period as signifying this fact. For example, to represent the position of the sun at a given instant, the astronomers of a certain century might accept all the values of right ascension, or all the values of declination, which do not differ from one another by more than one second, because their instruments do not permit them to distinguish points with an angular separation of less than one second from one another. The astronomers of the following century might have instruments with an optical power ten times greater. They might require then that the different determinations of the right ascension of the sun's center at a given moment, and the various determinations of the declination, do not differ from one another by more than a tenth of a second. An infinity of measurements, which would have satisfied their predecessors, would be rejected by them.

To the extent that the uncertainty of experimental results diminishes, the uncertainty of the formulas used to condense these results is going to diminish. One century accepted, as a law of motion for the sun, a whole group of formulas which gave the coordinates of the center of that star to within one minute at each instant. The following century will impose on any law of motion for the sun the condition that the coordinates of the sun's center be known to within a tenth of a second. An infinity of laws accepted by the first century thus will be rejected by the second.

Any physical law, being an approximate law, is at the mercy of progress that, by augmenting the precision of experiments, will render the degree of approximation of this law insufficient. Physicists must always consider it provisional.

It is not only because they are approximate that the laws of physics are provisional. It is also because they are symbolic relations. There are always cases in which their symbols are incapable of representing reality in a satisfactory manner.

To study a certain gas, physicists give a schematic representation of it. They represent it as a perfect fluid, having a certain density, raised to a certain temperature, and subject to a certain pressure. Among these three elements—density, temperature, and pressure—they establish a certain relationship: the law of compression and expansion of the gas. Is this law definitive?

Place the gas between the plates of a highly charged electric condenser. Determine its density, its temperature, and its pressure. The values of these three elements will no longer conform to the law of compression and expansion of the gas. Will physicists be surprised to find the law is inadequate? Will they place the fixity of the laws of nature in doubt? No. They will simply say that the defective relationship was a symbolic relationship and that it did not bear on the real gas that they manipulate, but on a certain schematic gas defined by its density, temperature, and pressure and that, no doubt, this schema was too simple and incomplete to represent the properties of the real gas placed in the conditions given now. Then they seek to complete the schema, making it more appropriate to express reality. They are no longer content to define the symbolic gas by means of its density, its temperature, and its pressure; they give it a dielectric capacity. In the representation of this body, they introduce the intensity of the electric field in which it is placed. They submit this more complete symbol to new studies and obtain the law of compressibility for a gas endowed with dielectric polarization. It is a more complicated law than the one they previously obtained. It includes the former as a particular case, but it is more comprehensive and will be verified in cases in which the latter fails.

Is this new law definitive?

Take the gas to which it applies, and place it between the poles of an electromagnet. Here the new law is refuted by experimentation in its turn. Do not think that this new refutation surprises physicists. They know that they have to deal with a symbolic relation and that the symbol they have created, a faithful image of reality in certain cases, might not resemble it in all circumstances. So they return to the schema of the gas on which they are experimenting without being discouraged. To permit this schema to represent the facts, they add some new properties. It is no longer enough that the gas has a density, a temperature, a dielectric capacity, a given pressure, and that it is placed in an electric field of a given intensity. They give it a coefficient of magnetization. They measure the intensity of the magnetic field in which it is placed, and, relating all these elements through a group of formulas, they obtain the law of compression and expansion of a polarized and magnetized gas. The law is more complicated but more comprehensive than those they have already obtained. The law will be verified in an infinity of cases in which the former will receive a refutation; nevertheless, it is a provisional law. One day, physicists foresee that conditions will be realized in which this law would find itself in error. On that day, it will be necessary to take up again the symbolic representation of the gas and add new elements to it. This symbol is like a mechanism whose flexibility increases with the number of pieces making it up, and which is more strictly applicable to the facts to the extent that it becomes more complex. But although this mechanism may well become more and more detailed and precise, it will always remain a rough and provisional model of reality.

This task of continual modification, through which the laws of physics better and better avoid experimental refutation, plays such an essential role in the development of physics that we may be permitted to insist somewhat more on its importance and to study its course in a second example.

Of all the laws of physics, the best verified by its innumerable consequences is surely the law of universal attraction. The most precise observations of the motions of the stars have not, until now, been able to show it in error. Is it a definitive law? Not at all, but a provisional law that must be modified and complicated endlessly to bring it into agreement with experimentation.

Here is some water in a vessel. The law of universal attraction tells you the force that acts on each of the particles of this water. This force is just the weight of the particle. Mechanics indicates to you what shape the water should assume. Whatever may be the nature and shape of the vessel, the water must be bounded by a horizontal plane. Look closely at the surface bounding this water. Although it is horizontal far from the edges of the vessel, the surface stops being horizontal in the vicinity of the glass wall. It rises along the length of these walls. In a confined space, it rises very high and becomes, to all intents, concave. Here the law of universal attraction is in error. To prevent capillary phenomena from refuting the law of gravitation, it must be modified. The inverse square law will have to be regarded as an approximate formula. We will have to admit that this formula expresses the attraction of two distant material particles known with sufficient precision, but that the formula becomes incorrect when it tries to express the attraction of two elements that are very close together. We will have to introduce an additional term in the equations, which, by complicating them, makes them appropriate to represent a more extended class of phenomena and permits them to encompass, in the same law, the movements of stars and capillary effects.

This law will be more comprehensive than Newton's law. For all that, it will not be exempt from all contradiction. In a mass of cases, the laws of capillarity will be in disagreement with the observations. In order to make this disagreement disappear, it will again be necessary to take up the formula for capillary action, modifying and completing it by taking account of the electric charges carried by particles of the fluid and of the forces acting between these electrified particles. Thus, the struggle will continue indefinitely between reality and the laws of physics. For any law formulated by physics, experimentation will oppose the brutal refutation of a fact. But physics will tirelessly retouch, modify, and complicate the refuted law, in order to replace it with a comprehensive law in which in turn the exception raised by the experiment will have found its rule.

Physics makes progress through this unceasing struggle, this effort to complete the laws and make them accommodate exceptions. It is because a piece of amber rubbed with silk placed the laws of gravity in error that physics created the laws of electrostatics. It is because a magnet produced effects contrary to these same laws of gravity that physics imagined the laws of magnetism. It is because Oersted found an exception to the laws of electrostatics and magnetism that Ampère invented the laws of electrodynamics and electromagnetism. Physics does not progress in the manner of geometry, which adds new definitive and indisputable propositions to the definitive and indisputable propositions already possessed. Physics progresses because experimentation endlessly causes new disagreements between theory and reality and because physicists endlessly retouch and modify theory to give it a better resemblance to reality.

IV. The laws of physics are more detailed than the laws of common sense

The laws that nonscientific experience permits us to formulate are general judgments whose meaning is immediate. In the presence of one of these judgments, we may ask: Is it true? In general, the answer is easy. The law is recognized as true for all times and without exception.

Scientific laws founded on experiments in physics are symbolic relations, and their meaning would remain unintelligible to anyone lacking knowledge of physical theories. Being symbolic, they are never either true or false. Like the experiments they rest on, they are approximate. Though sufficient today, the degree of approximation of a law will become insufficient in the future through the progress of experimental methods. Thus, a law of physics is always provisional. It is also provisional because it does not connect realities but symbols and because there are always cases in which the symbol no longer represents the reality. Hence, the laws of physics cannot be maintained except through a continual work of retouching and modification.

The problem of the certainty of physical laws presents itself in another manner, one more complicated and more delicate than the problem of the certainty of commonsense laws. We may be tempted to draw from this the strange conclusion that knowledge of physical laws constitutes a degree of science inferior to the simple knowledge of commonsense laws. To those who would seek to deduce this paradoxical consequence from the preceding considerations, we will content ourselves by repeating for the laws of physics what we have said for scientific experiments: A physical law possesses certainty that is much less immediate and much more difficult to appreciate than a commonsense law. But it surpasses the latter in the minute precision of its details.

"In Paris, the sun rises each day in the east, climbs into the sky, then descends and sets in the west." Let this commonsense law be compared with the formulas which at every moment reveal, to within one second, the coordinates of the center of the sun, and you will be convinced of the accuracy of the former proposition.

It is the care in minute exactness and precise analysis that distinguishes physical science from common sense. It is this carefulness that gives to its laws a provisional and approximate character. Everything we have just said about this character is, in some way, a commentary on this aphorism of Pascal: "Justice and truth are two points so fine that our instruments are too blunt to touch them exactly. If they do make contact, they blunt the point and press all round the false rather than the true."[9]

If, after that, people are still surprised to see the human mind, roughly informed about the phenomena of nature by laws that are certainly true, asking more detailed knowledge of these same phenomena from formulas that are only approximate and provisional, I will content myself by adding to their meditations the following apology:

9. [Pascal, *Pensées*, la Fuma ed., no. 44; Krailsheimer trans. (Middlesex: Penguin, 1966), p. 42.]

A botanist seeking a rare tree met two country people from whom he requested information. "There is one of those trees in this wood here," says the first. The other says to him, "Take the third path that you come to. Follow it for one hundred paces. You will be at the very foot of the tree you are seeking." The botanist takes the third path, he goes a hundred steps, but he does not reach the object of his quest. To touch the foot of the tree requires an additional five paces.

Of the two pieces of information that he received, the first was true and the second was false. Even so, which of the two country people has more right to his gratitude?

Conclusion

From these several reflections concerning the experimental method employed in physics, we may deduce many conclusions. I wish to emphasize only one of them.

Metaphysicians are accustomed, especially in our day, to borrow laws of physics in order to build up or destroy philosophical systems. The slightly superstitious faith that people in our time profess in the power and infallibility of positive science, and the reproach frequently and violently addressed to philosophers not to remain strangers to the conquests of this science both contribute to strengthening that tendency. That this tendency might be fundamentally legitimate is something that I will be careful not to deny. But, surely, it remains full of dangers for those who imprudently abandon themselves to it, and these are the dangers that I want to point out.

Let philosophers take care not to consider a law of physics as an absolute truth, partaking of the certainty of the mathematical propositions whose form it adopts. Physics knows no absolute truths. Although in making use of the language of mathematics physics shares its infallibility, it cannot state its laws in the language of algebra or geometry except on condition that they are regarded as approximate.

Let philosophers never regard a law of physics as an unbreakable and unlimited truth that will remain eternally true, that will never encounter exceptions. Laws are approximate, with an approximation that satisfies us, but which will no longer satisfy our successors. Every law of physics accepted today is destined to be rejected someday. Laws are symbolic. They apply not to reality but to an oversimplified schema. All physical laws are essentially provisional. They are always applied to an infinitely small number of cases in comparison to the number of cases that escape them.

They are unceasingly modified and made more complete to accommodate the facts which refute them, but they never exhaust the exceptions.

Above all, let philosophers not forget the symbolic character of physical laws. The magnitudes connected in the equations expressing these laws are no more than signs. In order to interpret these signs, we need a key, a complicated key made up of physical theories. Philosophers who wish to make use of the laws of physics must possess a profound understanding of these theories. Lacking this understanding, the significance that they attribute to these laws would be no more than nonsense.

5

Analysis of Mach's
The Science of Mechanics:
A Critical and Historical
Account of Its Development[1]

The following is Duhem's extended review of La Mécanique. Etude historique et critique de son développement, *from the fourth German ed. (Paris: Hermann, 1904), Emile Bertrand, French translator, by Ernst Mach, Emeritus Professor at the University of Vienna.*

Professor Mach's *Mechanics* offers some of the most varied reading that could be wished for. There one finds mathematical deductions, but as simplified as possible and devoid of useless displays of formulas; experiments, something surprising to a French reader in a treatise on mechanics, yet truly essential for understanding that science; some philosophy, but stripped of that pedantic jargon which thinks it has achieved depth when it plunges into obscurity; some historical pictures, but done in large strokes not laden by the minutiae of erudition; finally some polemics, but without bitterness or egotism. What was missing from this diversified, sober, and living book to seduce French readers? For it to be written in French. By translating this work, Emile Bertrand has taken from us any pretext for ignoring it any longer.

1. The work as a whole is dominated by a theory on the nature and scope of natural philosophy. We cannot discuss this theory in the *Bulletin*, since it is not a philosophy journal, but neither can we avoid giving a sum-

1. ["Analyse de l'ouvrage de Ernst Mach, *La mécanique. Etude historique et critique de son développement*," *Bulletin des science mathématiques*, 27 (1903): 261–283; also in *Revue des questions scientifiques* 55 (1904): 198–217.]

mary of it, for its consequences extend to the exposition of the rational mechanics presented by Professor Mach and impose on it its design.

The philosophical doctrine that gets Professor Mach to consider science as an *economy of thought* took form in him a long time ago. As early as 1868, he set it forth in a lecture on "The Forms of Liquids."[2] In 1882, he gave it a doctrinal formulation in a scholarly work.[3] Since then, he has pursued its consequences in various works, and notably in the one at present occupying us.

> My conception of economy of thought was developed out of my experience as a teacher, out of the work of practical instruction. I possessed this conception as early as 1861, when I began my lectures as Privat-Docent, and at the time believed that I was in exclusive possession of the principle—a conviction which will, I think, be found pardonable. I am now, on the contrary, convinced that at least some presentiment of this idea has always, and necessarily must have, been a common possession of all inquirers who have ever made the nature of scientific investigation the subject of their thoughts.[4]

From the time when Mach formulated his doctrine on the nature of natural philosophy, thoughts more or less similar to his have been developed in England, Germany, and France in the writings of numerous authors. Among these, some were subject more or less directly to the influence of the professor from Vienna. Others rediscovered these already discovered ideas by their own efforts without feeling the beneficial effects of his influence; naturally, they did not give the research of their unknown predecessor the acknowledgment it rightly deserved.[5]

The immense multitude, the infinite variety of the objects proposed to human knowledge would exceed immeasurably the extent and ability of human intellects if they needed to conserve in memory a simple copy of their personal experience. Furthermore, they would lack the time and

2. E. Mach, *Die Gestalten der Flüssigkeit* (Prag: Calve, 1872); E. Mach, *Populär-wissenschaftliche Vorlesungen* (Leipzig, 1896).

3. E. Mach, *Die ökonomische Natur der physikalischen Forschung* (Vienna: Gerold, 1882); E. Mach, *Populär-wissenschaftliche Vorlesungen* (Leipzig, 1896).

4. [Ernst Mach, *The Science of Mechanics: A Critical and Historical Account of Its Development*, trans. Thomas J. McCormack (1893; 6th ed., LaSalle, Ill.: Open Court, 1960), p. 591.]

5. Please allow us to excuse in this way the absence of the name of Mach from publications in which we have sometimes put forth thoughts that had more than mere similarity with his.

means for transmitting the fruits of this experience to contemporaries or to posterity. Therefore, before storing the contributions of perception in their minds, they must condense them, concentrate them, and extract their essence in such a way that they can lodge everything useful in that multitude of facts in as small a compartment as possible. This summation, this *abstraction* in the etymological sense of the word, is the proper object of scientific work. In every domain, the progress of science has as its aim to hold as much reality as possible in as reduced a form as possible; the essence of this progress is greater and greater *economy* of thought.

According to Mach, this tendency toward economy, considered as the directive principle of scientific labor, is particularly distinguishable in the domain of physical sciences.

In nature, there are only *facts*; the single law which alone will enter into science instead of and in the place of multiple facts is the copy, summed up by abstraction, of the characters common to all these facts—or, better, of those common characters that particularly interest us.

> In speaking of cause and effect we arbitrarily give relief to those elements to whose connection we have to attend in the reproduction of a fact in the respect in which it is important to us. There is no cause nor effect in nature; nature has but an individual existence; nature simply *is*. Recurrences of like cases in which *A* is always connected with *B*, that is, like results under like circumstances, that is again, the essence of the connection of cause and effect, exist but in the abstraction which we perform for the purpose of mentally reproducing the facts.[6]

Every physical law is therefore the economical summation of an immense number of facts; it allows us to know the character that, for one reason or another, we consider important, and not the whole of each of these facts.

> Thus, instead of noting individual cases of light refraction, we can mentally reconstruct all present and future cases, if we know that the incident ray, the refracted ray, and the perpendicular lie in the same plane and that $\sin \alpha / \sin \beta = n$. Here, instead of the numberless cases of refraction in different combinations of matter and under all different angles of incidence, we have simply to note the rule above stated and the values of *n*, which is much easier. The economical purpose is here unmistakable. In nature there is no *law* of refraction, only different cases of refraction. The law of refraction is a concise compendi-

6. [Mach, *The Science of Mechanics*, p. 580.]

ous rule, devised by us for the mental reconstruction in part, that is, on its geometrical side.[7]

The increasingly ample and general formulas of theoretical physics are only condensations, abstract summaries pushed further and further. "Science itself, therefore, may be regarded as a minimal problem, consisting of the completest possible presentation of facts with the *least possible expenditure of thought.*"[8]

2. These ideas on the nature of a formula of mechanics or physics direct the method that must serve to prove a similar formula. The latter formula claims to be only a condensed representation of experience. The only way of testing its validity, the only *demonstration* of which it must be susceptible, therefore consists in pitting it against the facts it wishes to represent; it will be better to the extent that it will represent a greater number of facts, with greater certainty, and by simpler proceedings.

> The function of science, as we take it, is to replace experience. Thus, on the one hand, science must remain in the province of experience, but, on the other, must hasten beyond it, constantly expecting confirmation, constantly expecting the reverse. Where neither confirmation nor refutation is possible, science is not concerned. . . . Those ideas that hold good throughout the widest domains of research and that supplement the greatest amount of experience, are the *most scientific.*[9]

It is therefore through the comparison of the set of consequences to the increasingly numerous facts of experience that the validity of a law can be established. But this process of demonstration is not a process of invention. It can be used only when the law is clearly formulated; it cannot suggest its formulation.

The processes of invention cannot be codified; the inventor of a law will allow the statement of the law to be suggested by the most varied considerations. Induction, generalization, and analogy will most often be the preferred guides.

But these factors, which push the inventor to consider such a proposition as the statement of a valid law, would not be enough to convince con-

7. [Mach, *The Science of Mechanics*, p. 582.]
8. [Mach, *The Science of Mechanics*, p. 586.]
9. [Mach, *The Science of Mechanics*, pp. 586–587.]

temporaries. On the other hand, the latter cannot test the validity of the law by submitting all its consequences to the control of facts; that test requires experiments that are not yet realized and which only the future of science can furnish. We are then led to give the new law a pretended *demonstration*.

Such a demonstration takes as axioms a certain number—the smallest possible—of propositions derived from our *instinctive knowledge*.

A prudent mind must keep itself on guard against the logical value of such demonstrations.

First, it is extremely difficult to enumerate all the instinctive knowledge which is really in play in such a deduction; almost no author succeeds in making all of it explicit without any omission or repetition.

Moreover, instinctive knowledge is, after all, only a confused and unanalyzed pile of experimental givens acquired at imprecise periods of intellectual development.

> [I]nstinctive knowledge enjoys our exceptional confidence. No longer knowing *how* we have acquired it, we cannot criticize the logic by which it was inferred. We have personally contributed nothing to its production. It confronts us with a force and irresistibleness foreign to the products of voluntary reflective experience. It appears to us as something free from subjectivity, and extraneous to us, although we have it constantly at hand so that it is more ours than are the individual facts of nature. All this has often led men to attribute knowledge of this kind to an entirely different source, namely, to view it as existing *a priori* in us (previous to all experience). . . . Yet even the authority of instinctive knowledge, however important it may be for actual processes of development, must ultimately give place to that of a clearly and deliberately observed principle. Instinctive knowledge is, after all, only experimental knowledge, and as such is liable, we have seen, to prove itself utterly insufficient and powerless, when some new region of experience is suddenly opened up. . . .
>
> It is more in keeping, furthermore, with the economy of thought and with the aesthetics of science, directly to *recognize* a principle (say that of the statical moments) as the key to the understanding of *all* the facts of a department, and *really see* how it *pervades* all those facts, rather than to hold ourselves obliged first to make a clumsy and lame deduction of it from unobvious propositions that involve the same principle but that happen to have become earlier familiar to us. . . . In fact, this mania for demonstration in science results in a rigor that is *false* and *mistaken*. Some propositions are held to be possessed of more certainty than others and even regarded as their necessary and incontestable foundation; whereas actually no higher, or perhaps not even so high, a degree of certainty attaches to them. Even the rendering clear of the degree of certainty

which exact science aims at, is not attained here. Examples of such mistaken rigor are to be found in almost every textbook.[10]

By studying the development of the principles of statics, Mach deploys all the resources of an unforgiving logic against this false rigor. This part of his book is perhaps the one that will cause the thinking reader to stop the longest.

3. How do we proceed, however, when we want to teach someone approaching a science such as mechanics one of these economical formulas that contain the concentrated and condensed essence of a number of facts? Will we forcefully express the relevant formula and limit ourselves to adding that the subsequent development of the theory will always show it to be in agreement with the facts? According to the preceding ideas, this method would be logical, but the most elementary psychology would show that it would be deplorable. Students would see only a form devoid of all content in the law presented in this fashion; it would remain unknown to them. How, then, can we prepare their minds to acquiesce to that proposition and to capture its sense? By representing a path similar to the one the inventor has followed; by examining the few facts the inventor has first studied; by reproducing the series of analyses and extensions by which the general law was derived. The real introduction to the expression of a principle of physics is a historical introduction:

> The fundamental elements of the notions that mechanics studies are almost completely developed by means of research on very simple special cases of mechanical phenomena. The historical analysis of these particular problems remains, in any case, the most efficacious and most natural means of penetrating the essential elements of the principles; one can even say that it is only by means of this path that it is possible to achieve a full comprehension of the general results of mechanics.

Lately, our teaching of physical science in secondary school tends more and more to reject historical considerations and to regard them as the object of empty and idle curiosity. Those who have tried to promote this tendency should meditate on the work of Professor Mach. I do not doubt that this reading would shake their conviction; it would contribute, I think, to giving them this completely opposite conviction, which the expe-

10. [Mach, *The Science of Mechanics*, pp. 93–94.]

rience of teaching or of examinations has brought to more than one professor: The person who does not know the erroneous principles replaced by a law of physics, at least in general, and the efforts it has taken to supplant them does not have a complete and penetrating comprehension of that law.

Moreover, Mach not only thinks that the study of history is of capital importance for understanding the science already accomplished, he also sees in it a precious guide for the inventor who wishes to open new paths:

> We shall recognize also that not only a knowledge of the ideas that have been accepted and cultivated by subsequent teachers is necessary for the historical understanding of a science, but also that the rejected and transient thoughts of the inquirers, nay even apparently erroneous notions, may be very important and very instructive. The historical investigation of the development of a science is most needful, lest the principles treasured up in it become a system of half-understood prescripts, or worse, a system of *prejudices*. Historical investigation not only promotes the understanding of that which now is, but also brings new possibilities before us, by showing that which exists to be in great measure *conventional* and *accidental*. From the higher point of view at which different paths of thought converge we may look about us with freer vision and discover routes before unknown.[11]

4. Whatever the importance Professor Mach attributes to the historical study of science, that study is for him a means, not an end. His object is not to revive for us the ideas of the first inquirers, to restore the first attempts at the doctrines that their successors have adopted, to follow in all its details the evolution by which these attempts have been organized, differentiated, and completed a little at a time, in order to become theories that are extended and detailed. He leaves these inquiries to the professional historian and to the psychologist. If he refers to history, it is only in order better to grasp the real and concrete meaning of the *economical* formulas that today constitute science.

The book we have before us does not pride itself on being a complete history of mechanics, one in which the progress of each of the branches of science is minutely followed from the appearance of the first bud to the maturity of the fruits. In the long series of transformations that constitute such progress, Mach has chosen only what helps us to understand the definitive plan—in the way that a zoologist would ask an embryologist to illuminate only the anatomy of the adult form. In making this choice, he abbreviates the exposition of the beginnings that are too ancient and too

11. [Mach, *The Science of Mechanics*, p. 316.]

confused for science to have kept their marks; he sets aside many attempts that have not succeeded, many seeds that have been aborted.

Mach fully justifies his right to choose, to attend only to certain phases of scientific development, and to treat history more as a logician than as a psychologist. His differences with Wohlwill on the subject of Galileo's ideas suggest the following to Mach:

> It is not to be denied that the different *phases* in the intellectual development of the great inquirers have much interest for the historian, and *some one* phase may, in its importance in this respect, be relegated into the background by the others. One must needs be a poor psychologist and have little knowledge of oneself not to know how difficult it is to liberate oneself from traditional views, and how even after that is done the remnants of the old ideas still hover in consciousness and are the cause of occasional backsliding even after the victory has been practically won. Galileo's experience cannot have been different. But with the physicist it is the instant in which a new view flashes forth that is of greatest interest, and it is this instant for which he will always seek. I have sought for it, I believe I have found it.[12]

Treated according to the method justified by Mach, the history of mechanics will appear infinitely interesting to the physicist, to whomever searches in the past only for lights capable of illuminating the present. If they forget that this is, in fact, the goal the author wished to reach, the historian and psychologist would no doubt address some objections to him.

They would criticize some important gaps in his exposition. The name of Descartes cannot be found in his history of statics; however, Descartes is the first to have clearly distinguished the two notions of force and of work, to have indicated the infinitesimal character of the principle of virtual displacements.

They would, above all, reproach Mach's historical pictures for being too simple, too clear, too perfectly ordered; the evolution retraced by these pictures tends too steadily, too surely, toward the goal it has to attain; in reality, the march of the human mind has been more hesitant, more tentative. It has strayed many times in the inextricable undergrowth of overly complex problems, and many times it has had to clear the brush at the edge of the precipice of an unfathomable question.

They would finally reproach history as conceived by Mach for being too *subjective*. It bears too deeply the mark of preoccupations that haunt the mind of the historian.

12. [Mach, *The Science of Mechanics*, p. 333.]

5. If we forgot that Mach prides himself on being a physicist and logician, rather than a historian, doubtless this last reproach would be addressed to him when reading the chapter devoted to "Theological, Animistic, and Mystical Points of View in Mechanics."

From the first lines of the Preface, the author presents his work as a "critical explanation animated by an *anti-metaphysical* spirit."

Today the foundations of theoretical mechanics and physics must be completely independent of any metaphysical system—*a fortiori* of any theological system. No one with sense, we believe, could contest the validity of that principle, which Mach formulates clearly and on several occasions.

But the general adherence of scientists to this principle is a wholly recent fact. If we proceed backwards, if we cast our eyes toward the past, we would recognize that, for a long time, mechanics and physics were bound most tightly with metaphysics, theology, and even occult sciences. To cite only one example, we would not be able to understand the objections raised against Newton's system by the atomists and Cartesians without returning to the scholastic metaphysical discussions on form and matter, quality and quantity. The very idea of universal attraction had its first roots nourished by astrological doctrines.

This constant action and reaction of philosophical and theological sciences on mechanics and physics must be constantly present to the mind of those who claim to resuscitate the ways of thinking of the creators of science. If they lost sight of them for only a second, they would quickly go astray in the midst of the discordances and debates under which the laws of natural philosophy have pursued their slow evolution.

But very often these laws, having achieved their definitive form, display themselves deprived of all the philosophical and theological ideas whose nourishment was for a long time necessary to their development. The adult no longer remembers the womb from which it was born.

Therefore, those who seek in the history of physical science only a more complete knowledge of its material and concrete content can almost always break the many links between this history and the history of philosophical and theological systems. Mach has kept only a few of these links. Some anecdotes show us that more than one of the creators of mechanics had religious faith. The list ends in the eighteenth century and terminates with Euler. The names of Ampère and Cauchy would have allowed its prolongation to the nineteenth century. We learn also that several of the great scientific innovators—among them Kepler—were not able to tear themselves away from the superstitions of their contemporaries. The links between science and philosophical ideas brought to our attention by Mach are the most fragile; they are not those who, by their vigorous and prolonged effort, have impressed a new and permanent direction on the march of mechanics. He should have rejected or ignored them so that his exposition would have lost nothing in depth and unity.

6. The *intellectual economy* that is the essential object of science, according to Professor Mach, attains its supreme degree in the form it gives to mechanics. In fact, he reduces this science to a single proposition, as follows:

> *Two parts of matter whose dimensions are very small determine accelerations on one another which are always directly opposed to one another; the magnitudes of these two accelerations always stand in a relation that is absolutely fixed for two given parts of matter.*

It is easy to see, however, that this statement is not in itself sufficient to constitute mechanics. At least two other propositions must be added, of which the first has already attracted the attention of the author. Here is that first proposition:

> *Let A, B, C be three small parts of matter; if only the pair BC is considered, the accelerations of B and of C are related among themselves by* α; *if the pair CA is considered, the accelerations of C and of A are related among themselves by* β; *finally, if the pair AB is isolated, the accelerations of A and of B are related among themselves by* γ. *The relation between the three numbers* α, β, γ *is*
> $$\alpha\beta\gamma = I$$

This relation alone allows us to attach an invariable number to each small part of matter—the *mass* of that particle—so that the relation of the mutual accelerations of the two particles is always equal to the inverse ratio of their masses.

The second proposition essential to the constitution of mechanics is the following:

> *In a system formed by a certain number of material particles, the acceleration of each of these particles can be regarded as the geometric resultant of accelerations, each of which is supposedly created by one of the other particles.*

The definition of mass is thus connected by Mach to the Newtonian law of the equality of action and reaction. This definition was first indicated by the author in a short paper entitled "Über die Definition der Masse." The idea was too novel; it was received very coldly. Poggendorff refused to publish it in his *Annales,* and it appeared more than a year later in Carl's *Repertorium der Experimental Physik.* Today it is accepted by a great number of those who teach mechanics.

7. According to the method proposed by Laplace, systematically followed by Poisson, and after him by a throng of students of mechanics, if we regard bodies as formed by small masses isolated from one another, the postulates proposed by Mach certainly suffice for writing the general equations of dynamics. Their sufficiency does not seem as certain or as evident to us if we wish, following the example of Lagrange, to treat bodies as continuous media whose various parts obstruct one another in their various motions and constitute *links* for one another.

Furthermore, Mach does not conceal his preferences for the method of Laplace and Poisson, which he calls the *Newtonian method*; in fact, we would not deny that this method is naturally linked to Newton's ideas. Here, for example, on the subject of the definition of solid bodies, are passages that Laplace and Poisson would not have disavowed:

> Nor, where a number of the masses m1, m2 . . . have considerable extension, so that it is impossible to speak of a *single* line joining every two masses, is the difficulty, in point of principle, any greater. We divide the masses into portions sufficiently small for our purpose, and draw the lines of junction mentioned between every two such portions. We, furthermore, take into account the reciprocal relation of the parts of the same large mass; which relation, in the case of rigid masses for instance, consists in the parts resisting every alteration of their distances from one another. On the alteration of the distance between any two parts of such a mass an acceleration is observed proportional to that alteration. Increased distances diminish, and diminished distances increase in consequence of this acceleration. By the displacement of the parts with respect to one another, the familiar forces of elasticity are aroused. When masses meet in impact, their forces of elasticity do not come into play until contact and an incipient alternation of form take place. . . .
>
> [N]o body is completely at rest, but . . . in all slight tremors and disturbances are constantly taking place which now give to the accelerations of descent and now to the accelerations of elasticity a slight preponderance. . . . The motion of an elastic body might in such case be characterized as vermicular. With hard bodies, however, the number of the oscillations is so great and their excursion so small that they remain unnoticed, and may be left out of account. . . . Here also in the case of sufficient hardness the vibrations may be

neglected. Bodies in which we purposely regard the mutual displacement of the parts as evanescent, are called *rigid* bodies.[13]

Mach gives the following conclusion to the argument from which we have just cited some extracts:

The considerations here developed will convince us that we can dispose by the Newtonian principles of every phenomenon of a mechanical kind which may arise, provided we only take the pains to enter far enough into details. We literally *see through* the cases of equilibrium and motion which here occur, and behold the masses actually impressed with the accelerations they determine in one another. It is the same grand fact, which we recognize in the most various phenomena, or at least can recognize in the most varied phenomena, or at least can recognize there if we make a point of so doing. Thus a unity, homogeneity, and economy of thought were produced, and a new and wide domain of physical conception opened which before Newton's time was unattainable.

Mechanics, however, is not altogether an end in itself; it has also *problems to solve* that touch the needs of practical life and affect the furtherance of other sciences. Those problems are now for the most part advantageously solved by other methods than the Newtonian—methods whose equivalence to that has already been demonstrated. It would, therefore, be mere impractical pedantry to contemn all other advantages and insist upon always going back to the elementary Newtonian idea. It is sufficient to have once convinced ourselves that this is always possible. Yet the Newtonian conceptions are certainly the most *satisfactory* and the most lucid; and Poinsot shows a noble sense of scientific clearness and simplicity in making these conceptions the sole foundation of the science.[14]

We do entirely accept these judgments of the professor from Vienna; we do not believe that there was always equivalence between the method of Lagrange and the method that Laplace and Poisson have derived from Newton's principles. We believe that the extreme intellectual economy that has presided over the constitution of this latter method has impoverished it too much for it to be able to furnish a satisfactory representation of all the phenomena of equilibrium and motion. But we have insisted sufficiently on these considerations elsewhere so that we may be allowed to abridge them here.[15]

13. [Mach, *The Science of Mechanics*, pp. 345–351.]
14. [Mach, *The Science of Mechanics*, p. 357.]
15. P. Duhem, *L'Évolution de la mécanique* (Paris, 1903), part I, chap. 8.

8. Let us now come to a singular difficulty raised by the principles of dynamics. Newton had already encountered this difficulty; many other thinkers have struggled with it after him. For about the last thirty years, Carl Neumann and Mach have again drawn the attention of philosophers and physicists to it.

First, it is certain that the *relative motion* of two bodies with respect to one another is the only motion that physicists can observe and about which geometers can reason. Both groups attribute a precise meaning to this proposition: The two bodies A and B move with respect to one another. It means that the set of the two bodies A and B does not have the same configuration at various instants of time. But let us not think about asking them whether it is body A that moves, or body B, or both at the same time; this question, as with every question concerning the *absolute motion* of a body, has no meaning for them. When they speak about the motion of a body, they always suppose that a choice has been made of a point of comparison, of a set of coordinate axes to which this motion is referred.

Second, it is true that students of mechanics cannot formulate the laws obeyed by the motions of a certain number of bodies, unless they are limited to considering the relative motions of these bodies. This remark is evident if one considers the fundamental proposition of dynamics as Mach formulates it. According to that proposition, if the two small determinate material parts are separated, their accelerations are directly opposed and the relation of these accelerations has an invariable value. Now, for someone who knows only the relative motion of the two particles considered, it is impossible to speak of the acceleration of each of these two particles; these words are devoid of meaning. The two particles have accelerations only if we assume that their combined motion is referred to a certain set of coordinate axes. But then these accelerations, their directions, and their relation will depend essentially on the coordinate axes that have been chosen. If the preceding proposition is correct after choosing a certain set of axes, it becomes false, in general, when we choose another, moving in an arbitrary motion with respect to the first.

The classic statement of the law of inertia would give rise to similar remarks. As this law states, an isolated material point continues in a rectilinear and uniform motion. For the geometer, as for the physicist, if a material point were alone in the world, it would be absurd to speak of its motion; what reference would allow this motion to be recognized? We cannot speak of the motion of a material point except by conceiving at the same time the existence of a point of comparison from which it is observed. But then, if the motion of that point is rectilinear and uniform when it is referred to a given point, it is no longer in general rectilinear and uniform when it is referred to another point in motion relative to the first.

These remarks that we could extend all lead to this conclusion: The fundamental statements of dynamics presuppose that all motions are referred to a single set of coordinate axes. If they are assumed to be correct with respect to a given set of axes, they will still be correct with respect to a second coordinate, provided the relative motion of these two axes is a uniform translation. Beyond this case, they would generally be false, if we refer the motions to a second coordinate.

We would have been able to develop observations similar to the ones we have just made concerning motion if we considered *time*. Neither the geometer nor the physicist would be able to talk about an *absolute time* but only of a *time relative* to a certain clock. All the statements of dynamics pre-suppose that a certain clock has been chosen. If they were true for a certain clock, they would no longer be, in general, for another clock, as long as the time marked by the latter were not a linear function of the time marked by the former.

Every mechanical system constructed according to the principles that Galileo, Huygens, and Newton formulated therefore presupposes the choice of a definite set of coordinate axes and definite clock. To this propo-sition one can add another which states no more than an approximate law: In agreement with experience, a simple mechanical system can be constructed by taking a set of coordinate axes that remain effectively linked to the fixed stars and a clock that makes diurnal motion sensibly uniform. Those who see in physical theories only a mathematical symbol capable of representing reality, but without natural relations to that reality, are easily contented with what has just been said. They admit without difficulty that in order to con-struct this symbol, an appeal must be made to a purely ideal set of coordi-nates and to a purely ideal clock. They are not shocked by the fact that nothing in reality corresponds to this set of coordinates or to this clock.

It is not the same for those who want to see in physical theories an instance and not a symbol of reality. The latter require that the point of comparison to which dynamics refers the motions and the clock on which it reads the time correspond, not just approximately but exactly, to real objects.

Some want these objects to be real, but they do not suppose them to be material. With Newton they admit the existence of an *absolute time* and an *absolute space*. They discuss the nature of this time and space as metaphy-sicians—like Clarke, who makes absolute time and absolute space the attributes of God.

Others, more positivistic, require the point to which mechanics refers the motions to have material existence. According to Carl Neumann, the form of the equations of mechanics postulates the existence of a certain body, the *absolutely fixed body* or α *body*. The existence of this body follows from the theories of dynamics and the verifications they find in experience, in the same way that the existence of the *aether* results from the success of wave optics. Budde, pushing the same conception further, believes that this α *body* is a medium in which other bodies reside. I have no objection to Budde's point of view [says Mach], but I think that the properties of this medium can be discovered by any physical process whatever and must not be accepted ad hoc. Today we do not have a sufficient notion of the properties of such a medium, or of the conditions of the motions of the bodies which reside in it.

9. All of dynamics is condensed into a very small number of propositions. If this dynamics accounted for all the phenomena that the world of matter presents to us, the *economy* of scientific thought would have reached its highest degree.

This *mechanical explanation* of all the phenomena of physics was long regarded as the proper object of science. Mach does not hesitate to regard this conception of physics as erroneous:

> The French encyclopaedists of the eighteenth century imagined they were not far from a final explanation of the world by physical and mechanical principles; Laplace even conceived a mind competent to foretell the progress of nature for all eternity, if but the masses, their positions, and initial velocities were given. In the eighteenth century, this joyful overestimation of the scope of the new physico-mechanical ideas is pardonable. Indeed, it is a refreshing, noble, and elevating spectacle; and we can deeply sympathize with this expression of intellectual joy, so unique in history.
>
> But now, after a century has elapsed, after our judgment has grown more sober, the world-conception of the encyclopaedists appears to us as a *mechanical mythology* in contrast to the *animistic mythology* of the old religions.[16]

Elsewhere, the professor from Vienna takes up the same idea. By condemning the excesses of mechanism, he defines the method according to which physics must henceforth progress. Let us cite in full the opening two pages of the chapter devoted to "The Relations of Mechanics to Physics." The thoughts expressed there seem to us correct and forceful:

16. [Mach, *The Science of Mechanics*, pp. 558–559.]

Purely mechanical phenomena do not exist. The production of mutual accelerations in masses is, to all appearances, a purely dynamical phenomenon. But with these dynamical results are always associated thermal, magnetic, electrical, and chemical phenomena, and the former are always modified in proportion as the latter are asserted. On the other hand, thermal, magnetic, electrical, and chemical conditions also can produce motions. Purely mechanical phenomena, accordingly, are abstractions, made, either intentionally or from necessity, for facilitating our comprehension of things. The same thing is true of the other classes of physical phenomena. Every event belongs, in a strict sense, to all the departments of physics, the latter being separated only by an artificial classification, which is partly conventional, partly physiological, and partly historical.

The view that makes mechanics the basis of the remaining branches of physics, and explains all physical phenomena by mechanical ideas, is in our judgment a prejudice. Knowledge which is historically first, is not necessarily the foundation of all that is subsequently gained. As more and more facts are discovered and classified, entirely new ideas of general scope can be formed. We have no means of knowing, as yet, which of the physical phenomena go *deepest*, whether the mechanical phenomena are perhaps not the most superficial of all, or whether all do not go *equally deep*. Even in mechanics we no longer regard the oldest law, the laws of the lever, as the foundation of all the other principles.

The mechanical theory of nature, is, undoubtedly, in an historical view, both intelligible and pardonable; and it may also, for a time, have been of much value. But, upon the whole, it is an artificial conception. Faithful adherence to the method that led the greatest investigators of nature, Galileo, Newton, Sadi Carnot, Faraday, and J. R. Mayer, to their great results, restricts physics to the expression of *actual facts*, and forbids the construction of hypotheses behind the facts, where nothing tangible and verifiable is found. If this is done, only the simple connection of the motions of masses, of changes of temperature, of changes in the values of the potential function, of chemical changes, and so forth is to be ascertained, and nothing is to be imagined along with these elements except the physical attributes or characteristics directly or indirectly given by observation.[17]

10. Mach has applied this method in various works to thermal phenomena;[18] from 1872 on, he gave some indications, reproduced in the present work, concerning the use that can be made of it in the study of electrical phenomena. The author justly remarks that the method he recommends is

17. [Mach, *The Science of Mechanics*, p. 597.]
18. E. Mach, *Die Geschichte und die Wurzel des Satzes der Erhaltung der Arbeit* (Prague, 1872); *Prinzipien der Wärmelehre* (Leipzig, 1896).

the one that Cohn and Hertz have used in some noted reports. It is fitting only to observe that Mach limited himself to treating electrostatic phenomena, while his successors grouped these phenomena and electromagnetic effects into the same theory.

Thermodynamics, as conceived by Kirchhoff and Mach, and electrical science, as constructed by Hertz and Cohn, are built on a plan similar to the one Mach imposes on mechanics. At the start, a small number of hypotheses and equations are directly *postulated* in all their generality. Thus, at the foundation of thermodynamics, we postulate the principle of the conservation of energy and the principle of Carnot-Clausius. At the foundation of electrical theory, we set down Maxwell's six equations and the expression of electrical energy. Mathematical analysis then derives a multitude of consequences from the postulated principles. Finally, these consequences are compared with the facts of experience. Concordance between the two is the proof that the theory is a good one.

As satisfactory as the theories so constructed are, they present a defect that does not allow them to satisfy the thinking person completely. They are *isolated* from one another. Each of them, issuing from autonomous principles, forms a chapter apart, with no link to the other chapters whose totality constitutes physics.

This parceling out of physical science would not be able to satisfy a philosopher convinced that "Every event belongs, in a strict sense, to all the departments of physics."[19] Thus, the professor from Vienna recommends the search for *analogical* links that could hold between the various parts of physics: "It is extremely useful to compare the directive concepts of the various domains of scientific knowledge among themselves."

This inquiry into the analogy between the various chapters of theory does not appear to him as the ultimate goal to attain, but as an advance toward a higher ideal:

> The pursuit of such resemblances and differences lays the foundation of a *comparative physics*, which shall ultimately render possible the concise expression of extensive groups of facts, without *arbitrary* additions. We shall then possess a homogeneous physics, unmingled with artificial atomic theories.[20]

The *economy of thought*, in which Mach sees the logical goal of science, pushes us, in fact, to substitute for ancient mechanics a science whose

19. [Mach, *The Science of Mechanics*, p. 596.]
20. [Mach, *The Science of Mechanics*, p. 599.]

increasingly general principles give us the summary representation of a set of increasingly numerous facts. For some years, we have seen attempts to erect such an *energetics* multiply.

Mach reproaches Hertz for having treated these attempts "more severely than is appropriate"; he himself speaks favorably of them. And even if he did not try to give us a project of energetics, we can count him among the precursors and promoters of this new doctrine; it is, in fact, the natural consequence of the principles he has postulated.

11. Mach is the resolute adversary of the philosophy, inaugurated by Descartes, that claims to reduce all the phenomena of the material world to motion. Moreover, the reaction he proposes against Cartesian philosophy goes farther. Cartesian philosophy had constructed a ditch as deep as an abyss between the *world of matter*, whose essence is *extension*, and the *world of mind*, whose essence is *thought*. Mach foresees the time when this ditch will be filled.

> Careful physical research will lead, however, to an analysis of our sensations. We shall then discover that our hunger is not so essentially different from the tendency of sulphuric acid for zinc, and our will not so greatly different from the pressure of a stone, as now appears. We shall again feel ourselves nearer nature, without its being necessary that we should resolve ourselves into a nebulous and mystical mass of molecules, or make nature a haunt of hobgoblins. The direction in which this enlightenment is to be looked for, as the result of long and painstaking research, can of course only be surmised. To *anticipate* the result, or even to attempt to introduce it into any scientific investigation of today, would be mythology, not science.[21]

The passage we have just cited would have been favorably received by Leibniz; for, according to him, the phenomena that bodies represent "do not consist only in bare extension and its change," but "something that has a relation with souls must necessarily be recognized." The passage would, above all, have been welcomed by the ancient scholastics as a return to their preferred doctrines; in fact, for them, as for Mach, the force that pulls the magnet toward the iron, the *alteration* engendered by the presence of the magnet in the *substantial form* of iron, was not essentially different from the sympathy or appetite which urges us toward a person or a thing, since this passion is nothing other than an *alteration* created on the soul by the presence of the object, the *substantial form* of man.

21. [Mach, *The Science of Mechanics*, p. 559.]

Mach is therefore subject, as are many others, to the effect of this great current that pushes scientific thought toward the doctrines we once thought had been abandoned forever.

Let us conclude this analysis, which is already too long for this *Bulletin* and too short to grasp the ample and numerous thoughts suggested by reading Mach's book. This book was written to prevent mechanics from degenerating into a series of correct and precise, but arid and sterile formulas. For reasons that are useless to enumerate, since the whole world knows them, in French teaching, mechanics has been reduced to a rigid dead form, emptied by degrees of all real content. In the introduction that he has written for the present work, E. Picard does not hesitate to describe the dynamics taught today as a "hierarchical and rigid science." Let teachers and students read and ponder over the *Mechanics* of Professor Mach. They will find in it the principles of a resurrection which will revive the living, throbbing flesh on the dried bones of that skeleton.

6

From *To Save the Phenomena*: Essay on the Concept of Physical Theory from Plato to Galileo[1]

These two chapters constitute the last two parts of a series of articles originally published in 1908 in Annales de Philosophie Chrétiennes *and collected into a monograph published as* SOZEIN TA PHAINOMENA *[To Save the Phenomena]. Duhem's introduction indicated the main thesis of the work, an extended essay supporting his interpretation of physical theory, as discussed in* The Aim and Structure of Physical Theory: *"What is the value of physical theory? What relations does it have with metaphysical explanation? These are lively questions today, but like many other questions, they are not new. They belong to all time. They have been raised ever since a science of nature has existed. The form in which they are cloaked may change a little from one century to another, because the variable form of these questions derives from the science of the day; but one need only remove this covering to recognize that the questions remain essentially the same.*

"Until the seventeenth century very few parts of natural science have progressed to the point of formulating theories in mathematical language, whose predictions, expressed numerically, can be compared with the measurements furnished by precise observations. Even statics, then called scientia de ponderibus, *and catoptrics, at that time ranked under perspective—our modern optics—had barely reached this degree of development. Setting aside these two restricted domains, we have before our eyes only one science whose form, already quite advanced, has anticipated the look of our modern theories of mathematical physics: that science is astronomy. Thus, where we say 'physical theory,' the*

---

1. [*SOZEIN TA PHAINOMENA: essai sur la notion de théorie physique de Platon à Galilée* (Paris: Hermann, 1908), chap. 7 and Conclusion, pp. 109–140.]

Greek or Arabic philosophers and the medieval or Renaissance savants said 'astronomy.'

"*The other parts of natural science had not yet reached that degree of perfection in which the language of mathematics serves to express laws discovered by precise observations. Positive physics, a science both mathematical and empirical, was not yet separated from the metaphysical study of the material world—from cosmology. In many instances, therefore, we would today speak of metaphysics, where the ancients used the word physics.*

"*That is why the question so much discussed today—what are the relations between physical theory and metaphysics?—was for two thousand years formulated in the following way: What are the relations between astronomy and physics?*"

From the Gregorian Reform of the Calendar to the Condemnation of Galileo

Astronomical hypotheses are simple devices for saving phenomena: As long as they reach that goal, they need not be true or even likely.

From the publication of Copernicus's book and Osiander's preface up to the Gregorian reform of the calendar, this opinion seems to have been generally accepted by astronomers and theologians. During the half century that follows, however, from the reform of the calendar to Galileo's condemnation, the opinion is relegated to oblivion; indeed, it is even violently attacked in the name of a universal realism. This realism sought assertions about the nature of things in astronomical hypotheses; thus, it required the hypotheses to be in agreement with doctrines of physics and with scriptural texts.

The learned Jesuit Christopher Clavius of Bamberg wrote a lengthy commentary on the *Sphaera* [*Sphere*] of John of Sacrobosco. The first two editions of that book, printed in Rome in 1570 and 1575, did not fully discuss astronomical hypotheses. In 1581, Clavius prepared a new edition "*multis ac variis locis locupletata* [enriched in many different places]."[2] On the back of the title page he enumerates the additions that enrich his third edition; among these is a "*disputatio perutilis de orbibus eccentricis et epicyclis contra nonnullos philosophos* [a useful disputation about eccentric and epicyclic orbs against several philosophers]."

2. Christopher Clavius, *In Sphaeram Ioannis de Sacro Bosco commentarius nunc iterum ab ipso Auctore recognitus, et multis ac variis locis locupletatus* (Rome, 1581).

This *disputatio*, entitled "*Eccentrici et epicycli quibus phenomena (phain-omenois) ab astronomi inventi sunt in coelo* [Eccentrics and epicycles, phenomena which have been devised by astronomers, are in the heavens]," is lengthy, taking up twenty-seven pages of very fine print.[3] It is also extremely interesting because it examines not only the Ptolemaic system but also the Copernican hypotheses. Clavius, in any case, was an admirer of the work of the astronomer of Thorn: When he mentions astronomical inventors, he cites his name several times; he mentions both the *De revolutionibus orbium coelestium* [*On the Revolutions of the Celestial Orbs*] and the *Tabulae prutenicae* [*Prutenic Tables*]; he goes so far as to call Copernicus "that most excellent geometer who, in our time, has restored astronomy, and who, in recognition of this, will be celebrated and admired as the equal of Ptolemy by all posterity." Such sentiments give a special weight to Clavius's critique of Copernican hypotheses.

Another circumstance increases the importance of these criticisms. As he tells us,[4] Clavius, a member of the Society of Jesus, was part of the commission instituted by Gregory XIII to prepare the reform of the calendar. He therefore seems to be an authoritative interpreter of the trends that prevailed in Rome at this time.

Clavius describes, only to reject, the opinion that treats eccentrics and epicycles as pure fictions devised solely to save the phenomena:

> Certain authors agree that all phenomena (*phainomena*) can be defended by assuming eccentric orbs and epicycles, but it does not follow, in their opinion, that these orbs really exist in nature. They are purely fictional; there may, in fact, exist another more convenient method for defending all the appearances, even though this method is unknown to us. Moreover, it may very well happen that the true appearances can be defended by the said orbs, even though they are entirely fictive and not at all the true causes of these appearances. For one can infer the true from the false, as Aristotle's dialectics shows.
>
> This argument can also be confirmed as follows: In his work entitled *De revolutionibus orbium coelestium*, Nicholas Copernicus saves all the phenomena (*phainomena*) in a different way. He assumes that the firmament is fixed and immobile. He also assumes that the sun is immobile at the center of the universe. As for the earth, he attributes to it a triple motion. Eccentrics and epicycles are therefore not necessary to save the phenomena (*phainomena*) of the wandering stars.[5]

3. Clavius, *Sphaera*, pp. 416–442.

4. Clavius, *Sphaera*, p. 61.

5. Clavius, *Sphaera*, pp. 435–436.

Clavius refuses to surrender to the force of these arguments. Of those who uphold them, he says:

> If they have a more convenient method, let them show it to us. We would be satisfied with it and greatly beholden to them. In effect, astronomers attempt solely to save all celestial phenomena (*phainomena*) in the most convenient manner, whether by means of eccentrics and epicycles or some other means. But since no more convenient method has been found until now than the one that saves all the appearances by means of eccentrics and epicycles, it stands to reason that the celestial spheres are constituted by orbs of that kind.

If one objected that the reality of some hypotheses cannot be proved from their agreement with phenomena, given that the impossibility of other hypotheses capable of saving the same appearances has not been established, Clavius would vigorously reject that objection; he would say that it would ruin all of physics, for physics is built entirely by proceeding from effects to causes. Sixty years earlier, Luiz Coronel had indicated the necessity to assimilate the theories of physics with astronomical hypotheses.

The fact, however, that Copernicus had succeeded in saving the appearances by means of a system distinct from Ptolemy's forces Clavius to attenuate his realist assertions—almost to reduce them to those Giuntini had formulated:

> It is not at all surprising that Copernicus should have succeeded in saving the phenomena (*phainomena*) in a different way. The motions of the eccentrics and epicycles has enabled him to know the time, magnitude, and quality of the appearances, future as well as past. Since he was extremely ingenious, he was able to imagine a new and, in his opinion, more convenient method, of saving the appearances. . . . Just as, when we know a correct conclusion, we can put together a series of syllogisms which derives it from false premises. But far from leading us to abandon eccentrics and epicycles, Copernicus's doctrine rather forces us to assume them. Astronomers have imagined such orbs because the phenomena have taught them in a manner more than certain that the wandering stars do not always remain at the same distance from the earth. . . . The only thing one can conclude from Copernicus's assumption is that it is not absolutely certain that the eccentrics and epicycles are arranged as Ptolemy thought, since a great number of phenomena (*phainomena*) can be defended by a different method. Now, in this question, we have only tried to persuade the reader that wandering stars do not always stay at one invariable distance from earth in their course; so there must be epicycles and eccentric orbs arranged in the heavens, as Ptolemy proposes, or, at least, some cause equivalent to eccentrics and epicycles must be placed there to account for these effects.[6]

6. Clavius, *Sphaera*, pp. 436–437.

This conclusion is almost, word for word, the cautiously formulated proposition of Guintini.

Copernicus's system provides precisely the causes which are equivalent to eccentrics and epicycles when accounting for astronomical phenomena. It therefore seems that, to conform to the rule he has just laid down, Clavius ought to have regarded Copernicus's theory to be as acceptable as Ptolemy's:

> If the assumption of Copernicus implied nothing false or absurd, one might be in doubt whether it is better to adhere to the opinion of Ptolemy or that of Copernicus, as long as it is a question of preserving the phenomena (*phainomena*). But Copernicus's position contains many absurd or erroneous assertions: it accepts that the earth is not at the center of the firmament; that it moves with a triple motion, something I cannot conceive, since according to the philosophers, a single simple body is entitled to one motion only; that the sun is at the center of the world and that it is devoid of motion. All these things are in conflict with the doctrine commonly received by philosophers and astronomers. Moreover, as we saw more fully in the first chapter,[7] these assertions seem to contradict what Holy Scriptures teach us in many places. That is why it appears to us that Ptolemy's opinion should be preferred over Copernicus's.
>
> From these considerations the following conclusion results: It is probable that there are eccentrics and epicycles; it is equally probable that there are eight or ten heavens, for it was by means of the phenomena (*phainomenois*) that astronomers have discovered this number of heavens and these orbits.

Clavius's position on astronomical hypotheses can be defined by means of the following propositions:

Astronomical hypotheses must save the phenomena as exactly and conveniently as possible, but this is not sufficient for them to be acceptable.

One cannot make certainty a condition of acceptability, but at least one can require that the astronomical hypothesis be probable.

To be probable, they must be compatible with the principles of physics; in addition, they must not contradict the teachings of the Church or scriptural texts.

---

7. Discussing the Copernican hypothesis of the motion of the earth in the first chapter, Clavius, defending the immobility of the globe, expressed himself as follows: "The Sacred Scriptures also support this opinion, for in many places they affirm that the earth is immobile, while the sun and other stars are in motion (*Favent huic quoque sententiae Sacrae Literae quae plurimis in locis Terram esse immobilem affirmant Solemque ac caetera astra moveri testantur*)." There followed a list of well-known texts. Clavius, *Sphaera*, p. 193.

Thus, two conditions of admissibility are imposed on any astronomical hypothesis that wishes to make its entry into science:

It may not be *falsa in Philosophia* [false according to philosophy].

It must be neither *erronea in Fide* [erroneous according to faith] nor, more important, *formaliter haeretica* [formally heretical].

These are the very criteria by which the Inquisition would judge the two fundamental hypotheses of the Copernican system in 1633. It was because both of them seemed *falsae in Philosophia* [false according to philosophy], one of them *ad minus erronea in fide* [at least erroneous according to faith], and the other *formaliter haeretica* [formally heretical], that the Holy Office prohibited Galileo from upholding them.

Three years before these two characteristics of any acceptable hypothesis were suggested in the work published in Rome by the Jesuit Christopher Clavius, they were formulated and applied at the other end of Europe by the Protestant Tycho Brahe.

Although it was not published until 1588,[8] by 1578, Tycho Brahe had completed the first eight chapters of his book on the comet of 1577.[9] Now, at the start of book 8, in order to propose a new theory, Brahe explains why he believes he must reject both the system of Ptolemy and that of Copernicus.[10]

By assuming that the rotation of the planet's deferent is uniform around the center of the equant, not around the center of that deferent, Ptolemy had adopted "hypotheses that violate the first principles of the art." Brahe therefore took into account "the innovation in the spirit of Aristarchus of Samos recently introduced by the great Copernicus"[11]:

> This innovation expertly and completely avoids everything superfluous or discordant in the system of Ptolemy. It violates no principles of mathematics. But it attributes to the earth, that coarse, lazy body, unfit for motion, a motion as

8. Tycho Brahe, *De mundi aetherei recentioribus phaenomenis liber secundus, qui est de illustri stella caudata anno 1577 conspecta* (Uraniborg, 1588). Our citation of this work follows the text reprinted in *Tychonis Brahe, mathim: eminent: Dani Opera omnia sive Astronomiae instauratae progymnasta in duas partes distributa, quorum* [sic] *prima de restitutione motuum Solis et Lunae, stellarumque inerrantium tractat. Secunda autem de mundi aetherei recentioribus phaenomenis agit* (Frankfurt, 1648).

9. Cf. Houzeau and Lancaster, *Bibliographie générale de l'astronomie*, vol. 1, p. 596.

10. Brahe, *De mundi aetherei*, pt. 2, p. 95.

11. Brahe, *De mundi aetherei*, pt. 2, p. 95.

quick as that of the ethereal fires and, what is more, a triple motion. In this way it stands refuted not only in the name of the principles of physics, but also in the name of the authority of the Sacred Scriptures. The latter, in fact, as we will show elsewhere more fully, several times affirm the immobility of the earth. . . .

Therefore, it has seemed to me that serious difficulties followed from both kinds of hypotheses (those of Ptolemy and those of Copernicus). Thus I meditated deeply, seeking to discover some hypothesis which would be rigorously established in all respects, both from the point of view of mathematics and the point of view of physics, one which would not be reduced to using subterfuges to avoid theological censure, and finally one which would fully accord with the celestial phenomena.

The principles posited by Osiander in his famous preface now looked to Tycho Brahe like mere subterfuges designed to avoid theological censure. Astronomical hypotheses should not only save the phenomena; they must also agree with both the principles of Peripatetic philosophy and Holy Writ, for they do not express mere fictions, but realities. However well the hypotheses of Copernicus are adapted to the appearances, they must be rejected because they cannot be brought into conformity with the nature of things. Tycho Brahe repeated this in the work published one year after his death through Kepler's efforts: "The arrangement which the great Copernicus attributed to the apparent circulations of the celestial bodies is extremely ingenious and well designed, but it does not, in reality, correspond to the truth."[12]

Brahe's opinions on the nature of astronomical hypotheses spread throughout Germany at the end of the sixteenth and the first years of the seventeenth century.[13]

We have before us the manuscript of a small treatise on astronomy, on the model of the *Sphaera* of Sacrobosco, composed in Wittenberg in 1604 by George Horst of Torgau.[14] Despite its elementary, textbook character—or, rather, by virtue of it—this small work is singularly appropriate for letting us know how astronomical hypotheses were viewed at the start

12. Tycho Brahe, *Astronomiae instauratae progymnasta, quorum haec prima pars de restitutione motuum Solis et Lunae stellarumque innerantium tractat* (Uraniborg, 1589; Prague, 1602), in Brahe, *Opera omnia*, pt. 1, p. 4.

13. [Duhem wrote: "durant la fin du XVe siècle et les premières années du XVIe siècle."]

14. George Horst, *Tractatus in arithmeticam Logisticam Wittebergaae privatim propositus . . . Introductio in Geometriam; Explicatio brevis ac perspicua doctrinae sphaericae in quatuor libris distributa* (1604).

of the seventeenth century at the celebrated Protestant university. It allows us to appreciate the extent of the changes in how this subject was seen in the fifty years since Melanchthon and Reinhold taught at that university. At the beginning of his little treatise, George Horst says:

> Astronomy is the science of motions that the heavenly bodies undergo, either in relation to one another, or in relation to the earth. It is called *scientia a potiori* [better science]; for although it shows only by sight (*katopsin*) some of the objects in the heavens, still it establishes most of its conclusions by means of apodictic principles, and it does this in a manner so certain and infallible that Pliny . . . rightly said: "it is shameful that anyone might doubt it."
>
> The principles of astronomy are of two kinds: true principles and analogical principles. The former are arithmetic and geometry. By means of these sciences, we raise ourselves to the sky, as if with wings, and in our flight we tour it in the company of the sun and the other stars. The latter are phenomena (*phainomena*) and hypotheses (*hupotheseis*). They are called analogical because they do not show that in virtue of which (*propter quid*) something exists or happens, but they demonstrate that something happens. . . .
>
> Everything that presents itself to observation through sight is called phenomena (*phainomena*).
>
> The *hypotheses* are assumptions made by the learned, assumptions by which they save and excuse the various phenomena (*phainomena*) produced in the sky. In this way, the man of science, who by nature desires to know the cause (*tou aitiou*), as Aristotle says in *Metaphysics*, book 1, comes to know the causes of these celestial changes and to reveal them to others. Among these hypotheses we find eccentric orbs, epicycles, and similar objects.

George Horst attributes absolute, apodictic certainty to these hypotheses, as he does to the phenomena. To ensure that nothing can cast any doubt on this certainty, he takes care to enumerate and to formulate with precision all the hypotheses he admits with respect to the sky, water, earth, etc. He appends to each hypothesis the reasons that guarantee its truth. These reasons are almost always arranged in two series: The author first enumerates those furnished by observation and peripatetic physics, and then those derived from scriptural texts.

The immobility of the earth, for example, is confirmed by two kinds of arguments, as it was in Melanchthon's *Initia Physicae*. But in invoking the two kinds of proofs in support of physical truth, Melanchthon left the astronomer free to save the phenomena by means of artificial hypotheses which were not in conformity with that truth. George Horst understands the hypotheses of astronomy as certain and infallible principles; that is why he tries to justify them by physical and theological arguments.

The adversaries of Copernicus's system came to rely more firmly on the principle that astronomical hypotheses express physical reality. It seems that this attitude might have pushed the Copernicans to take the opposite position—to maintain, with Osiander, that astronomical hypotheses are pure artifacts designed to save the phenomena. In fact, if they acknowledged that astronomical hypotheses must conform to the nature of things, they would be putting their system in great peril. On the one hand, their assumptions contradicted precisely those principles of peripatetic physics that most philosophers held to be certain, and they would be destroying those principles without proposing to replace them with anything; the hypothesis of the earth's motion, for example, was irreconcilable with scholastic teaching on the motion of projectiles, and no Copernican had attempted to provide a new theory of this motion. On the other hand, the motion of the earth and the immobility of the sun seemed formally denied by the Sacred Scriptures, and this objection must have appeared singularly strong to men who were for the most part sincere Christians, whether Catholic or Protestant.

Therefore, the Copernicans had every conceivable motive to lean toward the side recommended by the preface to the *De revolutionibus*. Yet the opposite side was the one they embraced. With considerably more passion than the Ptolemaists, they took it upon themselves to assert that astronomical hypotheses must be truths and that only Copernicus's assumptions conformed with reality.

Giordano Bruno is not merely passionate when, in one of his earliest writings, he rejects Osiander's opinion, attacking it most brutally.[15]

He reports that, according to some people, "Copernicus did not embrace the opinion that the earth is in motion, since it is an improper and impossible assumption; he attributed the motion to the earth instead of to the eighth sphere solely with a view toward the ease of calculation." But, says Bruno, the philosopher of Nola, "if Copernicus had affirmed the motion of the earth only for this cause and not for some other reason, it would seem to be a small thing, even an insignificant one. But it is certain that Copernicus believed in this motion, just as he affirmed it, and that he proved it with all his might." Bruno then speaks of "a certain preliminary epistle affixed to Copernicus's book by some ignorant and pre-

---

15. Michel di Castelnuovo, *La Cena de le ceneri. Descritta in cinque dialogi, per quattre interlocutori, contre considerationi, circa doi suggetti, all'unico refugio de le Muse* (1548). Reprinted in *Le Opere italiane di Giordano Bruno* (Göttingen: Paolo de Lagarde, 1888), vol. 1, pp. 150–152.

sumptuous ass who wanted, it seems, to excuse the author; or rather, even in this book, he wanted other asses to find the lettuce and small fruits he had left there so that they would not run the risk of leaving without lunch." Having presented Osiander's preface in such courteous terms, Giordano Bruno continues: "Behold the handsome doorman! See how good he is at opening the door and letting you enter to participate in this most honorable science without which the art of counting and measuring, geometry and perspective, would be no more than a pastime for ingenious madmen. Marvel at how faithfully he serves the master of the house!"

Although his sarcasm is in bad taste, Giordano Bruno was right when he denounced the contradiction between Osiander's preface and Copernicus's letter to Pope Paul III. He was right when he claimed that Copernicus "not only took on the office of the mathematician, who assumes the motion of the earth, but also that of the physicist, who demonstrates it." The realism professed by the Nolan philosopher about astronomical hypotheses is truly in the tradition of Copernicus and Rheticus.

Johann Kepler is, unquestionably, the strongest and most illustrious representative of that tradition.

Even in the preface of his first work, the *Mysterium cosmographicum*, printed in 1596, Kepler tells us that six years earlier, at Tübingen, as assistant to Michael Maestlin, he was already seduced by Copernicus's system: "[I] even wrote out a thorough disputation on the first motion, arguing that it comes about by the earth's revolution. I had then reached the point of ascribing to the same earth the motion of the sun, but where Copernicus did so through mathematical arguments, mine were physical, or rather metaphysical."[16]

Kepler was a Protestant but was deeply religious. He would not consider the hypotheses of Copernicus as conforming to reality if they were contradicted by Holy Scripture. Therefore, before advancing on the terrain of metaphysics or physics, he must first cross that of theology. At the beginning of chapter I of *Mysterium cosmographicum*, he tells us that "it is

16. Johann Kepler, *Prodromus dissertationum cosmographicarum continens mysterium cosmographicum de admirabili proportione orbium coelestium deque causis coelorum numeri, magnitudinis, motuumque periodicorum genuinis et propriis, demonstratum per quinque regularia corpora geometrica* (Tübingen: Georgius Gruppenbachius, 1596) in Kepler, *Opera*, ed. Frisch, vol. 1, p. 106. [Kepler, *Mysterium cosmographicum, The Secret of the Universe*, trans. A. M. Duncan (New York: Abaris, 1981), p. 63.]

proper to consider right from the start of this dissertation on Nature whether anything contrary to Holy Scripture is being said."[17]

Kepler thus indicates the path Copernicans will be required to follow from then on. As realists, they want hypotheses conforming to the nature of things; as Christians, they acknowledge the authority of the Sacred Text. They are, therefore, led to reconcile their astronomical doctrines with Scripture and forced to set themselves up as theologians.

If they had thought of astronomical hypotheses as Osiander did, they would have escaped this constraint. But those who faithfully followed the suggestions of Copernicus and Rheticus could not abide the doctrine expounded in the famous preface. Kepler says:

> I have never been able to agree with those who rely on the model of accidental proof, which infers a true conclusion from false premises by the logic of syllogism. Relying, I say, on this model they argued that it was possible for the hypotheses of Copernicus to be false and yet for the true phenomena (*phainomena*) to follow from them as if from authentic postulates. . . .
>
> Nor do I hesitate to affirm that everything which Copernicus inferred *a posteriori* and derived from observations, on the basis of geometric axioms, could be derived to the satisfaction of Aristotle, if he were alive [. . .], *a priori* without any evasions.[18]

As we saw earlier,[19] Nicolai Reymers [or Ursus] published his *De hypothesibus astronomicis* in 1597. There he took up again the doctrines that Osiander had expounded in the preface to *On the Revolutions*, on the subject of astronomical hypotheses. But if we judge by Kepler's analysis of Ursus's work,[20] the ideas of Copernicus's editor were disfigured by misleading exaggerations. We read there, for example, that "hypotheses are a fictive description of an imaginary form of the world system and not the real and true form of this system"[21]—an idea that Lefèvre d'Etaples developed magnificently. But we also read there that "hypotheses would not be hypotheses if they were true"; that "the proper object of hypotheses is to

17. Kepler, *Opera*, ed. Frisch, vol. 1, p. 112. [Kepler, *Mysterium cosmographicum, The Secret of the Universe*, p. 75.]

18. Kepler, *Opera*, ed. Frisch, vol. 1, pp. 112–113. [Kepler, *Mysterium cosmographicum, The Secret of the Universe*, pp. 75–77.]

19. See *To Save the Phenomena*, chap. 6.

20. We were unable to consult the work.

21. Kepler, *Opera*, ed. Frisch, vol. 1, p. 242.

derive the true from the false." The author of these assertions was playing with words; even if, in ordinary language, the word *hypothesis* has come to mean "doubtful assumption," philosophers and astronomers have kept its etymological meaning—that of a fundamental proposition upon which a theory rests.

To refute Ursus, Kepler composed a work, around 1600 or 1601, which was never completed and was not published until recently.[22] This essay has already given us important historical information about the preface that opens *De revolutionibus orbium coelestium*. It will now let us know Kepler's exact opinion concerning the nature of astronomical hypotheses:

> As in every discipline, as in astronomy also, the things we teach the reader by drawing conclusions we teach altogether seriously, not in jest. So we hold whatever there is in our conclusions to have been established as true. Besides, for the truth to be legitimately inferred, the premises of a syllogism, that is, the hypotheses, must be true. For only when both hypotheses are true in all respects and have been made to yield the conclusion by the rule of the syllogism shall we achieve our end—to reveal the truth to the reader. And if an error has crept into one or another of the premised hypotheses, even though the conclusion may occasionally be obtained, nonetheless, as I have already said in the first chapter of my *Mysterium cosmographicum*, this happens only by chance, and not always. . . . And just as in the proverb liars are cautioned to remember what they have said, so here false hypotheses, which together yield the truth once by chance, do not in the course of a demonstration in which they have been combined with many others retain that habit of yielding the truth, but betray themselves. . . . So since, as I have said earlier, none of those whom we honor as authors of hypotheses would wish to run the risk of error in his conclusions, it follows that none of them would knowingly admit among his hypotheses anything liable to error. Indeed they would worry not so much about the outcome and conclusions of demonstrations, but often more about the hypotheses they have adopted: thus all notable authors to date assess them on both geometrical and physical grounds and want them confirmed in all respects.[23]

Are there not distinct, though equivalent, hypotheses? Although they are incapable of being true simultaneously, do they not lead to identical

22. Kepler, *Apologia Tychonis contra Nicolaum Raymarum ursum*, in Kepler, *Opera*, ed. Frisch, vol. 1, p. 215.

23. Kepler, *Apologia Tychonis contra Nicolaum Raymarum ursum*, in Kepler, *Opera*, ed. Frisch, vol. 1, p. 239. [Translated by N. Jardine in *The Birth of History and Philosophy of Science* (Cambridge: Cambridge University Press, 1984), pp. 139–140.]

conclusions? Hipparchus's theorem, which allows solar motion to be represented equally well by an eccentric or by an epicycle rolling on a circle concentric with the world, provides a classic example. Is this not proof that true conclusions can be deduced from a hypothesis, though no astronomer could tell whether the hypothesis is or is not true?

In Kepler's opinion, this uncertainty is the lot of those astronomers who, in examining hypotheses, call only on mathematical reasons. The simultaneous use of reasons from geometry and reasons from physics will surely make it vanish:

> If a man assesses all things according to this precept, I doubt indeed whether he will come across any hypothesis, whether simple or complex, which will not turn out to have a conclusion peculiar to it and separate and different from all the others. Even if the conclusions of two hypotheses coincide in the geometrical realm, each hypothesis will have its own peculiar corollary in the physical realm. But practitioners are not always in the habit of taking account of that diversity in physical matters, and they themselves very often confine their own thinking within the bounds of geometry or astronomy and tackle the equivalence of hypotheses within one particular science, ignoring the diverse outcomes which dissolve and destroy the vaunted equivalence when one takes account of related sciences.[24]

The equivalence of two distinct hypotheses therefore can be only a partial equivalence. If certain conclusions can be equally deduced from two irreconcilable hypotheses, it is not by virtue of the differences between these two hypotheses but by virtue of what they have in common.

Here we encounter again the ideas of Adrastus of Aphrodisias and Theon of Smyrna.

Kepler is not content merely to criticize the doctrine maintained by Osiander and Ursus. He intends, further, to practice the realism whose principles he has proposed. The most considerable monument his genius has erected, the *Epitome Astronomiae Copernicanae*, bears witness to this realism.

Realism is affirmed from the beginning of the first book: "Astronomy," Kepler says, "is a portion of physics."[25] And the importance of this aphorism is shown immediately in what the author tells us about the causes of hypotheses (*de causis hypothesium*):

24. [N. Jardine, *The Birth of History and Philosophy of Science*, pp. 141–142.]

25. Kepler, *Epitome Astronomiae Copernicanae usitata forma quaestionum et responsionum conscripta, inque VIII libros digesta, quorum hi tres priores sunt de doctrina physica* (Lenz: Johannes Plancus, 1614), in Kepler, *Opera*, ed. Frisch, vol. 6, p. 119.

The third part of the astronomer's baggage is physics. In general, it is not considered necessary for the astronomer; and yet the astronomer's science has a great bearing on the object of this portion of philosophy, which, without the astronomer, would not reach completion. Astronomers should not, in fact, be given absolute license to assume anything whatever without sufficient reason. You ought to be able to give probable reasons for the hypotheses you claim to be the true causes of appearances. You ought, therefore, at the start, to seek the foundations of your astronomy in a higher science, I mean, in physics or metaphysics. Then, sustained by these geometric, physical, or metaphysical arguments that your particular science has provided, you will not be prohibited from leaving the limits of that science in order to discourse about the things that pertain to these higher doctrines.[26]

In the course of the *Epitome*, Kepler does not miss an occasion to support his hypotheses with arguments furnished by physics and metaphysics. What physics and what metaphysics! But this is hardly the place to tell what strange reveries, what childish fancies Kepler designated by these two words. We do not wish to investigate how Kepler constructed his astronomy; it suffices for us to know how he wanted it constructed. And, as we now know, he wanted the science of the celestial motions to rest on foundations guaranteed by physics and metaphysics, and he required that astronomical hypotheses not be contradicted by the Scriptures.

In addition, a new ambition is asserted in Kepler's writings: Once astronomy is founded on true hypotheses, its conclusions will be able to contribute to the advancement of the physics and metaphysics that initially supplied its principles.

At first, Galileo adopted the hypotheses of Ptolemy. In 1656, a small treatise on cosmography by the great Pisan geometer was printed in Rome;[27] this treatise was included in the second volume of the Padua edition of Galileo's works, published at Padua in 1744.[28] A brief note by the editor indicates the existence of a manuscript copy of the same work. According to this manuscript copy, Galileo wrote the work in 1606 to serve as a textbook for the students at Padua. Later editions of Galileo's works reproduce this brief treatise.

26. Kepler, *Opera*, ed. Frisch, vol. 6, pp. 120–121.

27. Galileo Galilei, *Trattato della sfera o Cosmografia* (Rome, 1656).

28. Galileo Galilei, *Opere divise in quattro tomi, in questa nova edizione accresciute di molte cose inedite* (Padua, 1744), vol. 2, p. 514.

It is extremely interesting to compare Galileo's treatise with the *Expositio doctrinae sphaericae* of George Horst, written two years earlier in Wittenberg. There are great similarities between motives of the two authors. Like Horst, Galileo first speaks of the various factors that go into the composition of astronomy. He points out phenomena, then hypotheses. Like Horst, he defines hypotheses: "Certain assumptions about the structure of celestial orbs that correspond to the appearances"; Galileo continues: "Since we are now at the first principles of this science, we will set aside the more difficult calculations and demonstrations, and deal solely with hypotheses. We will concentrate on confirming them and establishing them by means of the appearances."

How does Galileo conceive this confirmation of hypotheses? Will he ask them only to *save the phenomena*, or will he require that they be true or at least likely? Like Horst, he will not be satisfied with so little. He also wants the foundations of astronomical theory to conform to reality; similarly, he claims to demonstrate them by means of the classical proofs of Scholastic physics. There is only one notable difference between Galileo's demonstrations and those of Horst: Whenever he can, the Protestant professor of Wittenberg University adds the force of scriptural texts to the reasons derived from Aristotle's physics. The Catholic professor at the University of Padua never appeals to those texts.

When Galileo embraced Copernicus's system, he did so in the same spirit that had inspired him when he held Ptolemy's system. He wanted the hypotheses of the new system to be not artifacts for the calculation of tables but propositions in conformity with the nature of things. He wanted them to be established through physical reasons. One might even say that this physical confirmation of the Copernican hypotheses is the center toward which Galileo's most diverse investigations converge. His observations as an astronomer and his theories of mechanics even converge toward this goal. Furthermore, since he wanted the foundations of Copernican theory to be true, and since he did not think that the truth could contradict the divinely inspired Scriptures, he was led to reconcile his assertions with biblical texts. In time he turned theologian, as his famous letter to Grand-Duchess Christina of Lorraine testifies.

In claiming that the hypotheses express physical truths, and in declaring that they do not seem to him to contradict Holy Writ, Galileo, like Kepler, was plainly in the tradition of Copernicus and Rheticus. He collided with those who represented the tradition of Tycho Brahe the Protestant and Christopher Clavius the Jesuit. What they had said around the year 1580 the theologians of the Holy Office solemnly proclaimed in 1616.

The Holy Office seized on these two fundamental hypotheses of the Copernican system:

Sol est centrum mundi et omnino immobilis motu locali. Terra non est centrum mundi nec immobilis, sed secundum se totam movetur, etiam motu diurno. [The sun is the center of the world and never moves by local motion. The earth is not the center of the world and not immobile, but is moved as a whole by a diurnal motion.]

They asked themselves whether these two propositions bore the two marks which, by a common accord, Copernicans and Ptolemaists required of any admissible astronomical hypothesis: Are these propositions compatible with sound physics? Are they reconcilable with divinely inspired Scriptures?

For the Inquisitors, sound physics was the physics of Aristotle and Averroës. It dictated a straightforward reply to the first question: The two suspect hypotheses were *stultae et absurdae in Philosophia* [foolish and absurd philosophically].

As for the Scriptures, the advisors to the Holy Office refused to accept any interpretation that did not have the authority of the Church fathers on its side. Hence the reply to the second question was inescapable: The first proposition was *formaliter haeretica* [formally heretical], the second *ad minus in fide erronea* [at least erroneous in faith].

The two censured propositions did not present either of the two characteristics that marked an admissible astronomical hypothesis; therefore, both must be totally rejected, not to be used, even for the sole purpose of *saving the phenomena*. Thus Galileo was prohibited from teaching Copernicus's doctrine *in any way*.

The condemnation brought forth by the Holy Office resulted from the clash of two realist positions. This violent collision might have been avoided—the debate between the Ptolemaists and the Copernicans might have been kept to the terrain of astronomy alone—if certain wise precepts concerning the nature of scientific theories and the hypotheses on which they rest had been heard. These precepts, formulated by Posidonius, Ptolemy, Proclus, and Simplicius, had come down directly to Osiander, Reinhold, and Melanchthon by an uninterrupted tradition. But now they seemed forgotten.

There were, however, some voices determined to have them heard once again.

One of these voices was Cardinal Bellarmine, the same person who, in 1616, examined the Copernican writings of Galileo and Foscarini. As early as April 12, 1615, Bellarmine had written Foscarini a letter full of wisdom and prudence. Here are some passages from it:

> It seems to me that Your Eminence and Signore Galileo are proceeding prudently by limiting yourselves and speaking suppositionally (*ex suppositione*) and not absolutely, as I have always believed that Copernicus spoke. For there is no danger in saying that, by assuming the earth moves and the sun stands still, one saves all the appearances better than by postulating eccentrics and epicycles; and that is sufficient for the mathematician. However, it is different to want to affirm that in reality the sun is at the center of the world and only turns on itself without moving from east to west, and the earth is in the third heaven and revolves with great speed around the sun; this is a very dangerous thing, likely not only to irritate all scholastic philosophers and theologians, but also to harm the Holy Faith by rendering Holy Scripture false. . . .
>
> If there were a true demonstration that the sun is at the center of the world and the earth in the third heaven, and that the sun does not circle the earth but the earth circles the sun, then one would have to proceed with great care in explaining the Scriptures. . . . But I will not believe that there is such a demonstration, until it is shown me. Nor is it the same to demonstrate that by supposing the sun to be at the center and the earth in heavens one can save the appearances, and to demonstrate that in truth the sun is at the center and the earth in heaven; for I believe that the first demonstration may be available, but I have very great doubts about the second, and in a case of doubt one must not abandon the Holy Scripture as interpreted by the Holy Fathers.[29]

Galileo knew of the letter addressed by Cardinal Bellarmine to Foscarini; several writings published between the time when he learned of the letter and his first condemnation contained rebuttals to the cardinal's arguments. Reading these letters (whose excerpts Berti was the first to publish) enables us to capture vividly the spirit of Galileo's thought about astronomical hypotheses.

One piece, written toward the end of 1615 and addressed to the consultants at the Holy Office, warns them against two errors: The first is to claim the mobility of the earth to be, in some way, *an immense paradox and obvious foolishness* that has not yet been demonstrated and never will be

29. This letter was published for the first time in Domenico Berti, *Copernico e le vicende del sistema copernicano in Italia nella seconda meta del secolo XVI e nella prima del secolo XVII* (Rome, 1876), pp. 121–125. [Maurice A. Finocchiaro, *The Galileo Affair* (Berkeley: University of California Press, 1989), pp. 67–68.]

demonstrated; the second is to believe that Copernicus and the other astronomers who proposed this mobility "did not believe that it was true in fact and in nature" but admitted it only suppositionally, insofar as it can account more conveniently for the appearances of celestial motions and facilitate astronomical calculations.[30]

In affirming that Copernicus believed in the reality of the hypotheses formulated in *De revolutionibus* and in proving, by an analysis of the work, that Copernicus did not admit the earth's mobility and the sun's immobility only *ex suppositione*, as Osiander and Bellarmine wanted it, Galileo was maintaining the historical truth. But what interests us more than his judgment as historian is his opinion as physicist. Now this is easily conjectured from the piece we are analyzing: Galileo thought that the reality of the earth's motion was not only demonstrable but already demonstrated.

This thought stands out still more clearly in another text; in it not only do we see that Galileo thought the Copernican hypotheses could be demonstrated, but we also learn how he understood the demonstration to have been carried out:

> Not to believe that there is a demonstration of the earth's mobility until it is shown is very prudent, nor do we ask that anyone believe such a thing without demonstration. On the contrary, we only seek that, for the advantage of the Holy Church, one examine with the utmost severity what the followers of this doctrine know and can advance, and that nothing be granted them unless the strength of their arguments greatly exceeds that of the reasons for the other side. Now, if they are not more than ninety percent right, they may be dismissed; but if all that is produced by philosophers and astronomers on the opposite side is shown to be mostly false and wholly inconsequential, then the other side should not be disparaged, nor deemed paradoxical, so as to think that it could never be clearly proved. It is proper to make such a generous offer since it is clear that those who hold the false side cannot have in their favor any valid reason or experiment, whereas it is necessary that all things agree and correspond with the true side.
>
> It is true that it is not the same to show that one can save the appearances with the earth's motion and the sun's stability, and to demonstrate that these hypotheses are really true in nature. But it is equally true, or even more so, that one cannot account for such appearances with the other commonly accepted system. The latter is undoubtedly false, while it is clear that the former, which

---

30. Berti, *Copernico e le vicende del sistema copernicano in Italia nella seconda meta del secolo XVI e nella prima del secolo XVII*, pp. 132–133. [Finocchiaro, *The Galileo Affair*, p. 70.]

can account for them, may be true. Nor can one or should one seek any greater truth in a position than that it corresponds with all particular appearances.[31]

If we pressed this last proposition somewhat, we might easily make it yield Osiander's doctrine, which Bellarmine upheld—namely, precisely the one Galileo is attacking. Thus, logic constrains the great Pisan geometer to formulate a conclusion directly contrary to the one he had hoped to establish. But in the preceding lines, his thought stood out quite clearly.

The impending debate appears to his mind's eye as a sort of duel. Two doctrines announce themselves, each claiming to be in possession of the truth; but one speaks truly and the other lies. Who will decide? Experience. The one doctrine it refuses to agree with will be recognized as erroneous, and, by the same fact, the other doctrine will be proclaimed to conform with reality. The destruction of one of the two opposing systems assures the certainty of the opposite system, just as, in geometry, the absurdity of a proposition entails the correctness of its contradictory.

If anyone doubted that Galileo really held the opinion we are attributing to him about the proof of an astronomical system, they will be convinced, we believe, by reading the following lines:

> The quickest and surest way to show that Copernicus's position is not contrary to Scriptures is, as I see it, to show by a thousand proofs that this proposition is true and that the contrary position cannot subsist in any way; consequently, since two truths cannot contradict each other, it is necessary that the position recognized as true agree with the Holy Scriptures.[32]

Galileo's notions of the value of the experimental method and the art of using it are nearly those that Bacon was later to formulate. Galileo conceives of the proof of a hypothesis along the lines of reduction to absurdity

31. Berti, *Copernico e le vicende del sistema copernicano in Italia nella seconda meta del secolo XVI e nella prima del secolo XVII*, pp. 1329–1330. [Finocchiaro, *The Galileo Affair*, p. 85.]

32. Berti, *Copernico e le vicende del sistema copernicano in Italia nella seconda meta del secolo XVI e nella prima del secolo XVII*, pp. 105–106. [Berti's footnote: Codice Volpicelliano A. DCLVII. V. Vedi intorno a questo e gli altri Codici volpicelliani Documenti e note illustrative XI: "Il modo per me speditissimo e sicurissimo per provare che la posizione Copernicana non è contraria alle scritture sarebbe il monstrare con mille prove che ella è vera e che la contraria non non può in modo alcuno sussistere onde non potendo le verità contrariarsi, è necessario che quella e le Sacre Scritture sieno concordissime."]

proofs in geometry. Experience, by convicting one system of error, confers certainty on its opposite. Positive science progresses by a series of dilemmas, each resolved with the help of a crucial experiment (*experimentum crucis*).

This manner of conceiving of experimental method was bound to become extremely fashionable because it is very simple; but it is completely false because it is too simple. Grant that the phenomena are no longer saved by Ptolemy's system; Ptolemy's system will certainly be recognized as false. But it does not in any way follow from this that Copernicus's system is true, because Copernicus's system is not purely and simply the contradictory of Ptolemy's system. Grant that Copernicus's hypotheses succeed in saving all the known appearances. It can be concluded that these hypotheses may be true; but it cannot be concluded that they are certainly true. To justify this last proposition would require a proof that no other set of hypotheses could be imagined that allow the appearances to be saved just as well. And this last demonstration has never been given. Indeed, in Galileo's own time, were not all the observations that could be invoked in favor of Copernicus's system just as well saved by that of Tycho Brahe?

These remarks had often been made before Galileo's time. Their correctness exploded in front of the Greeks the day Hipparchus succeeded in saving the sun's motion equally by an eccentric or by an epicycle. Thomas Aquinas had formulated them with the greatest clarity. Nifo, Osiander, Alessandro Piccolomini, and Giuntini had repeated them. Once again, an authoritative voice was to remind the illustrious Pisan of them.

Cardinal Maffeo Barberini, who was soon to be elevated to the papacy under the name of Urban VIII, met with Galileo after the condemnation of 1616 to discuss the Copernican doctrine. Cardinal Oregio, present at this meeting, left us an account of it. At this meeting, the future pope, by means of arguments similar to those just rehearsed, laid bare the hidden error of Galileo's argument that since celestial phenomena all agree with Copernicus's hypotheses and are not saved by Ptolemy's system, Copernicus's hypotheses are certainly true and, moreover, necessarily in agreement with Holy Scriptures.

According to Oregio's account, the future Urban VIII advised Galileo

> to note carefully whether or not there is agreement between the Holy Scriptures and what he had conceived about the earth's motion, with an eye to saving the phenomena manifested in heaven and what philosophers commonly regard as acquired by observation and minute examination concerning the motions of

heaven and the stars. Having agreed with all the arguments presented by that most learned man, [Barberini] asked if God would have had the power and wisdom to arrange differently the orbs and stars in such a way as to save all the phenomena that appear in heaven or that refer to the motion, order, location, distances, and arrangement of the stars.

If you deny this [. . .], then you must prove that for things to happen otherwise than you have presented implies a contradiction. In fact, God in his infinite power can do anything which does not imply a contradiction; and since God's knowledge is not inferior to his power, if we admit that he could have done so, then we have to affirm that he would have known how.

And if God had the power and knowledge to arrange these things otherwise than has been presented, while saving all that has been said, then we must not bind divine power and wisdom in this manner.

Having heard these arguments, that most learned man was quieted.[33]

The man who was to become Urban VIII had clearly reminded Galileo of this truth: No matter how numerous and precise are experimental confirmations, they can never render a hypothesis certain, for this would require, in addition, demonstration of the proposition that these same experimental facts would forcibly contradict all other imaginable hypotheses.

Did these logical and prudent admonitions of Bellarmine and Urban VIII convince Galileo, sway him from his exaggerated confidence in the scope of experimental method, and in the value of astronomical hypotheses? We may well doubt it. In his celebrated *Dialogue* of 1632 on the two chief world systems, Galileo declares from time to time that he treats Copernicus's doctrine as a pure astronomical hypothesis without claiming it to be true in nature. But these protestations are belied by the accumulation of proofs of his interlocutor, Salviati, in favor of the reality of the Copernican positions; they are undoubtedly mere pretexts for getting around the prohibition of 1616. At the very moment when the dialogue is about to end, Simplicio, the stubborn and narrow-minded peripatetic on whom the thankless task of defending Ptolemy's system falls, concludes with the words:

33. Oregio, *Ad suos in universas theologiae partes tractatus philosophicum praeludium completens quatuor tractatus . . .* (Rome, 1637), p. 119. The same account is found on p. 194 of Oregio's *De Deo uno*, written in 1629 (Cf. Berti, *Copernico e le vicende del sistema copernicano in Italia nella seconda meta del secolo XVI e nella prima del secolo XVII*, pp. 138–139). [Trans. in Maurice A. Finocchiaro, *Galileo and the Art of Reasoning* (Dordrecht: Reidel, 1980), p. 10.]

I admit that your thoughts seem to me to be more ingenious than many others I have heard. I do not therefore consider them true and conclusive; indeed, keeping always before my mind's eye a most solid doctrine that I once heard from a most eminent and learned person, and before which one must fall silent, I know that if asked whether God in his infinite power and wisdom could have conferred upon the watery element its observed reciprocating motion using some other means than moving its containing vessels, both of you would reply that he could have. . . . From this I forthwith conclude that, this being so, it would be excessive boldness for anyone to limit and restrict the divine power and wisdom to some particular fancy of its own.

Salviati replies:

An admirable and angelic doctrine, and well in accord with another one, also divine, which, while it grants to us the right to argue about the constitution of the universe . . . adds that we shall not discover the work of his hands.[34]

Through the mouth of Simplicio and Salviati, Galileo perhaps wished to address a delicate piece of flattery to the pope. Perhaps he also wanted to answer the argument of Cardinal Maffeo Barberini with a joke; that is how Urban VIII took it. He gave free reign to the intransigent realism of the peripatetics of the Holy Office against the impenitent realism of Galileo. The condemnation of 1633 was to confirm the verdict of 1616.

Conclusion

Many philosophers since Giordano Bruno have reproached Osiander harshly for the preface he placed at the beginning of Copernicus's book. The advice given to Galileo by Cardinal Bellarmine and Urban VIII has been treated with hardly less severity since its publication.

The physicists of our day have weighed the value of the hypotheses used in astronomy and physics more minutely than did their predecessors. They have seen many illusions dispelled that previously passed for certainties, and they have now been compelled to acknowledge and proclaim that logic was on the side of Osiander, Bellarmine, and Urban VIII, not on the side of Kepler and Galileo. The former had understood the exact scope of the experimental method, and, in this respect, the latter were mistaken.

34. [Galileo, *Dialogue Concerning the Two Chief World Systems*, trans. Stillman Drake (2nd ed., Berkeley: University of California Press, 1970), p. 464.]

Yet the history of the sciences celebrates Kepler and Galileo and ranks them as the great reformers of experimental method, whereas Osiander, Bellarmine, and Urban VIII are passed over in silence. Is this history's supreme injustice? Could it not be, on the other hand, that those who attributed a false scope to experimental method and who exaggerated its value worked harder and better toward perfecting it than did those whose valuation was measured more precisely and correctly from the start?

The Copernicans stubbornly stuck to an illogical realism, even though everything drove them to give up that error, even though by attributing to astronomical hypotheses the right value as determined by so many authoritative voices, they could easily have avoided both the quarrels of philosophers and the censure of theologians. This strange conduct demands an explanation. How can it be explained except by the lure of some great truth, a truth too vaguely perceived by the Copernicans for them to be able to formulate it in its purity, to disengage it from the errors concealing it, yet a truth sensed so vividly that neither the precepts of logic nor the advice of self-interest could diminish its invisible attraction. What was this truth? This is what we want to try to articulate.

Throughout antiquity and the Middle Ages, physics displayed two parts so distinct from each other as to be, so to speak, opposed to each other: On the one hand, there is the physics of celestial and imperishable things; on the other, the physics of sublunary things subject to generation and corruption.

The objects treated by the first of these two physics are considered to be infinitely higher in nature than are those of the second; hence, it is concluded that the former is incomparably more difficult than the latter. Proclus teaches that sublunary physics is accessible to humans, whereas celestial physics is beyond them and is reserved for the divine intellect. Maimonides shares Proclus's opinion: According to him, celestial physics is full of mysteries whose knowledge is reserved for God's understanding, whereas physics, fully constituted, is found in Aristotle's work.

Contrary to what the men of antiquity and the Middle Ages thought, the celestial physics they had constructed was conspicuously more advanced than their terrestrial physics.

Since the time of Plato and Aristotle, the science of the stars was organized on the plan we still impose today on the study of nature. On the one hand was astronomy: geometers such as Eudoxus and Calippus constructed mathematical theories by means of which celestial motions could be described and predicted, whereas observers judged the degree of agreement between the predictions resulting from calculations and natural phenomena. On the other hand was physics, properly speaking—or, in modern terms, celestial cosmology: Thinkers such as Plato and Aristotle meditated on the nature of the stars and the cause of their motions. What relations held between these two parts of celestial physics? What precise boundary separated them? What affinity united the hypotheses of the one with the conclusions of the other? These are questions discussed by astronomers and physicists during antiquity and the Middle Ages and resolved in different ways, for their minds were directed by different motives, very similar to the motives that attract modern thinkers.

It happens that the physics of sublunary things reached this comparable degree of differentiation and organization in its own time. In modern times, it also divided into two parts, similar to those into which celestial physics has been divided since antiquity. The theoretical part constructs mathematical systems that give knowledge of the exact laws of phenomena by means of their formulas; the cosmological part seeks to fathom the nature of bodies, their attributes, the forces to which they are subject or which they exert, and the combinations they can acquire from one another.

In ancient times, during the Middle Ages and in the Renaissance, it would have been difficult to make this division. Sublunary physics hardly knew mathematical theories. Only two parts of that physics, optics (or *perspectiva*) and statics (or *scientia de ponderibus*), had taken on that form, and physicists were at a loss when they wanted to assign *perspectiva* and *scientia de ponderibus* their proper places in the hierarchy of the sciences. Aside from these two parts, analysis of the laws presiding over phenomena remained rather inexact and purely qualitative; it was not yet freed from cosmology.

In dynamics, for example, the laws of free fall, glimpsed since the fourteenth century, and the laws of projectile motion, vaguely surmised in the sixteenth century, continued to be involved in metaphysical discussions about local motion, natural and violent motion, and the coexistence of the source of motion and a moving object. Not until the time of Galileo do we see the theoretical part of physics become disengaged from the cosmological part at the same time that its mathematical form is being articulated. Until then, these two parts remained intimately united—or, rather, inextricably entangled. Their aggregate constituted the physics of local motion.

Meanwhile, the ancient distinction between the physics of celestial bodies and the physics of sublunary things was gradually obliterated. After Nicholas of Cusa and Leonardo da Vinci, Copernicus had dared to assimilate the earth to the planets. Tycho Brahe, through the study of the star that appeared and then disappeared in 1572, had shown that the stars themselves can be generated and corrupted. Galileo, by discovering sunspots and mountains on the moon, had completed the reunion of the two physics into a single science.

Henceforth, when the likes of Copernicus, Kepler, and Galileo declared with one voice that astronomy should take as hypotheses only propositions whose truth was established by physics, this seemingly single assertion contained in fact two distinct propositions.

In fact, such an assertion can mean that the hypotheses of astronomy are judgments about the nature of celestial objects and their real motions, that the experimental method, by serving to control the correctness of astronomical hypotheses, will come to enrich our cosmological knowledge with new truths. This first meaning lies, so to speak, at the very surface of the assertion. It became evident first. It is the sense the great astronomers of the sixteenth and seventeenth centuries saw clearly, the one they expressed formally, and the one that ultimately attracted their support. Yet given this meaning, their assertion is false and harmful. Osiander, Bellarmine, and Urban VIII rightly viewed it as contrary to logic. The assertion had to engender countless misunderstandings in human science before it was finally rejected.

Beneath this first, illogical but visible and seductive meaning, the assertion of the Renaissance astronomers contained another meaning. The requirement that the hypotheses of astronomy agree with the teachings of physics was a requirement that the theory of celestial motions rest on bases that were equally capable of supporting the theory of the motions observed here below. The path of the stars, the ebb and flow of the sea, projectile motion, and the fall of heavy bodies were required to be saved by the same set of postulates formulated in the language of mathematics. And this meaning remained deeply hidden. Neither Copernicus nor Kepler nor Galileo perceived it clearly; it remained disguised, yet fertile, beneath the clear but false and dangerous meaning the astronomers alone grasped. And while the false and illogical meaning they attributed to their principle engendered disputes and quarrels, the true but hidden meaning of this same principle gave birth to the scientific efforts of those inventors. While straining to maintain the truth of the former meaning, they were, unwittingly, establishing the correctness of the latter. When Kepler tried again and again to account for the motions of the stars by means of the properties of water currents or magnets, and when Galileo tried to make the path of projectiles agree with the motion of the earth or to derive an explanation of the tides from this motion, both believed they were proving that the

Copernican hypotheses had their foundation in the nature of things. But the truth they were introducing little by little into science was that a single dynamics must represent the motion of the stars, the oscillations of the sea, and the fall of heavy bodies, by means of a single set of mathematical formulas. They thought they were renovating Aristotle; they were preparing for Newton.

Despite Kepler and Galileo, we believe today, with Osiander and Bellarmine, that the hypotheses of physics are only mathematical artifacts devised for *saving the phenomena*. But thanks to Kepler and Galileo, we now require that they *save at the same time all the phenomena* of the inanimate universe.

7

Letter to Father Bulliot, on Science and Religion[1]

Duhem's outspokenness on issues that mattered to him, regardless of his target, can be seen clearly in a letter to his mother about his participation at the Third International Congress of Catholics, held in Brussels in 1894: "Thus, yesterday, I decided to strike a great blow. It was in the session on philosophy. The room was full, especially with clerics. A brave churchman just treated an objection taken from mechanics. My opinion was solicited concerning the scientific aspects of the lecture. Then, I said frankly to all these good Catholic philosophers that, as long as they continued to talk about science without knowing a single word of it, the freethinkers would have fun at their expense, that, in order to talk about questions relating to science and Catholic philosophy, one needs to have ten or fifteen years of pure science, and that, as long as they have not trained people with deep scientific knowledge, they must keep silent. . . . The idea was launched; it made its way. The whole afternoon nothing else was spoken of during the congress. I do not regret having come. I believe that the seed I have sowed will germinate. It is the first time these brave people heard the truth spoken. It surprises them a bit, but I am also surprised to see that they, or at least several of them, go about it with good will."[2] At the same congress, Duhem met the future head of the Department of Philosophy of the Institut Catholique in Paris, the Père J. Bulliot. Some years later, in 1911, Duhem wrote to Bulliot on the relations of science and religion. The letter allowed

1. [Hélène Pierre-Duhem, *Un Savant français, Pierre Duhem* (Paris: Plon, 1936), pp. 158–169.]

2. [Hélène Pierre-Duhem, *Un Savant français, Pierre Duhem*, p. 157.]

Duhem to synthesize his thoughts on the philosophy of science and the history of science in ways that he had not previously accomplished.

Bordeaux, 21 May, 1911

My Father,

I have heard that the Catholic Institute of Paris is preparing to organize a coordinated set of philosophical classes. This news caused me great joy, and I think it will cause great joy for all enlightened Catholics. It is time, in fact, that we oppose the numerous and learned teachings of indifferent or opposed philosophies with a whole college of chairs in which traditional Catholic philosophy is taught in all its power and in its full development.

I have some thoughts on the subject of the composition of the future Institute of Philosophy; I ask your permission to relate them to you. They are not advice. Coming from me, that would be impertinent. They are, rather, simple information. Living among those who profess doctrines contrary to ours, I am well placed to understand their plan of attack against us and to see where our defenses must be reinforced, above all.

The field on which the battle has been waged and where, without any doubt it will become more violent, is the incompatibility of the scientific mind and the religious mind.

I do not say: the incompatibility of a given scientific discovery with a given religious doctrine. The polemics of the nineteenth century were constituted of these particular antagonisms. For example, some used their ingenuity to oppose a given geological theory with a given verse of the Bible. But these were isolated sorties that prepared for the great fight. The latter is much more extensive; the result toward which it leads threatens to be much more radical. It concerns denying all religion the right to subsist, and that in the name of the whole of science. It is claimed, as established, that no sensible person could accept the validity of science and believe in the dogmas of religion at the same time. And since the validity of science is further affirmed every day by a thousand marvelously useful inventions, which only a blind soul could put into doubt, religious faith is done for.

Science, it is said, takes as foundation either axioms that reason cannot doubt or facts that have all the certainty of the testimony of the senses. Everything it erects on these foundations is constructed with the help of rigorous reasoning. And with an overabundance of precautions, experience controls each of the conclusions at which it arrives. The whole edifice therefore maintains the unshakable solidity of the first foundations.

Religious dogmas, on the other hand, issue from vague aspirations and intuitions arising from sentiment and not from reason. They are not subject to any rule of logic and cannot sustain the examination of any critique, however rigorous, even for an instant.

Hence, everything that had made up the object of religious dogmas will be declared absurd and devoid of sense, and we will be content with a strict and absolute positivism, a relative of that crass materialism, which comes like a forced conclusion. Or else, we will regard an object that escapes the demonstrations of science as incapable of being known with even the least certainty. We will profess an agnosticism according to which religion is only a more or less poetic and consoling dream. But how can people who have experienced the firm realities of science allow themselves to be lulled by such a dream?

They are not satisfied with placing into evidence, with the help of logic, this antagonism between the scientific mind and the religious mind. They also want the history of the development of human knowledge to make this evident for less enlightened people. They allegedly show that all sciences are born from the fertile Hellenic philosophy, whose most brilliant members abandoned the ridiculous task of believing in religious dogmas to the common people. They then depict this terrifying night of the Middle Ages during which the schools, enslaved to the conduct of Christianity and uniquely solicitous of theological discussions, were not able to glean even the least part of the Greek scientific heritage. They display the brilliance of the Renaissance in which minds finally liberated from the yoke of the Church rediscovered the thread of scientific tradition at the same time as the secret of artistic and literary beauty. They are happy to oppose the always ascendant progress of science and the always deeper decline of religion, beginning with the sixteenth century. They then believe we are allowed to predict the approaching death of religion at the same time as the universal and incontestable triumph of science.

That is what is taught from a mass of chairs and what is written in a multitude of books.

It is time that Catholic teaching address this teaching and that it respond to its adversaries with the following word: "lies!" These are lies in the domain of logic and lies in the domain of history. The teaching that claims to establish the irreducible antagonism between the scientific mind and the Christian mind is the most colossal, boldest lie that has ever attempted to dupe the human race.

In order to oppose the method that leads to scientific truths and the method that leads to religious dogmas, they falsify the description of both these methods. They consider both of them in a superficial manner and, as it were, externally. They borrow some features from this quick examination and pretend that these features are the very essence of the processes they claim to have analyzed.

How different are these methods to a mind that has really penetrated to the heart of the matter and that has captured their living principles! That mind is able to recognize both what gives these processes their variety and what unifies them. It sees everywhere a single human reason utilize the same essential means in order to arrive at the truth. But in each domain, it sees reason adapt the use it makes of these means to the special object whose knowledge it wishes to acquire. Thus, with the help of common operations that properly constitute our intellect, it sees the pursuit of a method for mathematical sciences, a method for physics, a method for chemistry, one for biology, one for sociology, and one for history. For mathematics, physics, chemistry, biology, sociology, and history have different principles and different objects. In order to reach those objects, we must follow, in the same fashion, different routes from different points of departure. It then recognizes that in order to attain religious truths, human reason uses no other means than those it used to reach the other truths. But it uses them in a different manner, because the principles from which it departs and the conclusions toward which it tends are different. The antagonism announced between scientific demonstration and religious intuition disappears while it perceives the harmonious agreement of the multiple doctrines by which our human reason forces itself to express truths of different orders.

What can be said about the strange history which we claimed was almost confirmed by our insufficient logical analysis?

From its birth, Greek science was fully impregnated with theology, but with a Pagan theology. This theology teaches that the heavens and the stars are gods. It teaches that they have no other motion than circular and uniform motion, which is perfect motion. It damns the impiety that would dare to attribute a motion to the earth, the sacred hearth of divinity. Although these theological doctrines furnished natural science with some temporarily useful postulates, although they guided its first steps, they soon became for physics what apron strings become for a child—namely, fetters. If the human mind did not break out of these fetters, it could not have surpassed Aristotle in physics or Ptolemy in astronomy.

Now, how did it break these fetters? The answer is Christianity. Who first profited from this freedom just gained in order to rush toward the discovery of a new science? Scholasticism. Thus, who dared to declare, in the middle of the fourteenth century, that the heavens were not moved by divine or angelic intelligences but by an indestructible impulse given by God at the moment of creation, in the manner of a ball thrown by a player? A Parisian Master of Arts, Jean Buridan. Who, in 1377, declared the diurnal motion of the earth simpler and more satisfying for the mind than the diurnal motion of the heaven? Who cleanly refuted all the objections raised against the first of these motions? Another Parisian Master, Nicole Oresme. Who founded dynamics, discovered the laws of falling bodies, postulated the foundations of geology? Parisian scholasticism, at a time when the Catholic orthodoxy of the Sorbonne was proverbial throughout the world. What role did the highly exalted freethinkers of the Renaissance play in the formation of modern science? In their superstitious and rote admiration of antiquity, they mistook and disdained all the fertile ideas put forth by fourteenth-century scholasticism, in order to take up again the least-sustainable theories of Platonic or peripatetic physics. What did this great intellectual movement amount to, which, at the end of the sixteenth century and the beginning of the seventeenth century, produced the doctrines accepted ever since? A pure and simple return to the teaching of Parisian scholasticism during the Middle Ages, so that Copernicus and Galileo are the heirs and, as it were, disciples of Nicole Oresme and Jean Buridan. Therefore, if this science of which we are justly proud has been able to see the light of day, it is because the Catholic Church has been its midwife.

Such are the denials we must issue against the false assertions found everywhere, in history as in logic. Do you not believe, my father, that this would be one of the most important roles, perhaps even the most important role that the future Institute of Philosophy will have to play? That is why I tend to think that two chairs would be particularly appropriate in this institute. The one devoted to the analysis of the logical methods by which the various sciences progress would show us that we could, without contradiction or incoherence, pursue the acquisition of positive knowledge and at the same time meditate on religious truths. The other, following the development of human science in its path, would lead us to recognize that when people cared most of all about the kingdom of God and his justice, God gave them in addition the deepest and most fruitful thoughts about the things here below.

Would you judge me very bold to have thus communicated my opinions to you? Surely not, for you know that the only thing that guides me in this affair is the desire to see God's rule reestablished among us. For such an aim, boldness is not only permitted, it is commanded.

Moreover, at this moment, when in face of the intellectual anarchy facing the human mind, I cry out to God: "*Adveniat regnum tuum* (Thy kingdom come)," I seem to hear your prayer echoing mine. May our prayers be granted! That is my wish in offering you my very humble respects.

Pierre Duhem

8

History of Physics[1]

This article attempts to give a complete picture of how the history of physics fits together. Readers who follow these issues into the contemporary history of science will find many qualifications to Duhem's various historical theses. On the crucial issue of the indebtedness of seventeenth-century physics to fourteenth-century physics, however, recent scholarship has tended to support Duhem's view. The article is also a good source of historical information for the other essays in the collection.

I. A Glance at Ancient Physics

Although at the time of Christ's birth, Hellenic science had produced nearly all its masterpieces, it was still to give to the world Ptolemy's astronomy, the way for which had been paved for more than a century by the works of Hipparchus. The revelations of Greek thought on the nature of the exterior world ended with the *Almagest*, which appeared about A.D. 145, and then began the decline of ancient learning. Those of its works that escaped the fires kindled by Islamic warriors were subjected to the barren interpretations of Muslim commentators and, like parched seed, awaited the time when Latin Christianity would furnish a favorable soil in which they could once more flourish and bring forth fruit.[2] Hence it is that the

1. [*Catholic Encyclopaedia* (New York: R. Appleton, 1911) vol.12, pp. 47–67. Lightly edited.]

2. [This article contains many statements that might be corrected in the light of modern scholarship. We have chosen to let them stand because of the historical

time when Ptolemy put the finishing touches to his *Great Mathematical Syntax of Astronomy* seems the most opportune in which to study the field of ancient physics. An impassable frontier separated this field into two regions in which different laws prevailed. From the moon's orbit to the sphere enclosing the World extended the region of beings exempt from generation, change, and death, of perfect, divine beings, and these were the star-sphere and the stars themselves. Inside the lunar orbit lay the region of generation and corruption, where the four elements and the mixed bodies generated by their mutual combinations were subject to perpetual change.

The science of the stars was dominated by a principle formulated by Plato and the Pythagoreans, according to which all the phenomena presented to us by the heavenly bodies must be accounted for by combinations of circular and uniform motions. Moreover, Plato declared that these circular motions were reducible to the rotation of solid globes all limited by spherical surfaces concentric with the World and the Earth, and some of these homocentric spheres carried fixed or wandering stars. Eudoxus of Cnidus, Calippus, and Aristotle vied with one another in striving to advance this theory of homocentric spheres, its fundamental hypothesis being incorporated in Aristotle's *Physics* and *Metaphysics*. The astronomy of homocentric spheres could not, however, explain all celestial phenomena, a considerable number of which showed that the wandering stars did not always remain at an equal distance from the Earth. Heraclides Ponticus in Plato's time and Aristarchus of Samos about 280 B.C. endeavored to account for all astronomical phenomena by a heliocentric system, which was an outline of the Copernican mechanics; but the arguments of physics and the precepts of theology proclaiming the Earth's immobility readily obtained ascendancy over this doctrine, which existed in a mere outline. Then the labors of Apollonius of Perga (at Alexandria, 205 B.C.), of Hipparchus (who made observations at Rhodes in 128 and 127 B.C.), and finally of Ptolemy (Claudius Ptolemaeus of Pelusium) constituted a new astronomical system that claimed that the Earth was immovable in the center of the universe; a system that seemed, as it were, to reach its com-

---

interest of the article. However, Duhem's dismissal of Islamic science, here and elsewhere, cannot pass without challenge. For a radically different view, see A. I Sabra (1987), "The Appropriation and Subsequent Naturalization of Greek Science in Medieval Islam," *History of Science* 25: 223–243; and F. Jamil Ragep (1990), "Duhem, the Arabs, and the History of Cosmology," *Synthèse* 83: 201–214.]

pletion when, between A.D. 142 and 146, Ptolemy wrote a work called *Megale mathematike syntaxis tes astronomias*, its Arabic title being transliterated by the Christians of the Middle Ages, who named it *Almagest*. The astronomy of the *Almagest* explained all astronomical phenomena with a precision which for a long time seemed satisfactory, accounting for them by combinations of circular motions; but of the circles described, some were eccentric to the World while others were epicyclic, the centers of which described deferent circles concentric with or eccentric to the World; moreover, the motion on the deferent was no longer uniform, seeming so only when viewed from the center of the equant. Briefly, in order to construct a kinematical arrangement by means of which phenomena could be accurately represented, the astronomers whose work Ptolemy completed had to set at naught the properties ascribed to the celestial substance by Aristotle's *Physics*, and between this *Physics* and the astronomy of eccentrics and epicycles there ensued a violent struggle which lasted until the middle of the sixteenth century.

In Ptolemy's time, the physics of celestial motion was far more advanced than the physics of sublunary bodies, as, in this latter science of beings subject to generation and corruption, only two chapters had reached any degree of perfection—namely, those on optics (called perspective) and statics. The law of reflection was known as early as the time of Euclid, about 320 B.C., and to this geometer was attributed, though probably erroneously, a *Treatise on Mirrors*, in which the principles of catoptrics were correctly set forth. Dioptrics, being more difficult, was developed less rapidly. Ptolemy already knew that the angle of refraction is not proportional to the angle of incidence, and in order to determine the ratio between the two, he undertook experiments, the results of which were remarkably exact.

Statics reached a fuller development than optics. The *Mechanical Questions* ascribed to Aristotle were a first attempt to organize that science, and they contained a kind of outline of the principle of virtual velocities, destined to justify the law of the equilibrium of the lever; besides, they embodied the happy idea of referring to the lever theory the theory of all simple machines. An elaboration, in which Euclid seems to have had some part, brought statics to the stage of development in which it was found by Archimedes (about 287–212 B.C.), who was to raise it to a still higher degree of perfection. It will here suffice to mention the works of genius in which the great Syracusan treated the equilibrium of the weights suspended from the two arms of a lever, the search for the center of gravity, and the equilibrium of liquids and floating bodies. The treatises of

Archimedes were too scholarly to be widely read by the mechanicians who succeeded this geometer; these men preferred easier and more practical writings such as, for instance, those on the lines of Aristotle's *Mechanical Questions*. Various treatises by Heron of Alexandria have preserved for us the type of these decadent works.

II. Science and Early Christian Scholars

Shortly after the death of Ptolemy, Christian science took root at Alexandria with Origen (about 180–253), and a fragment of his *Commentaries on Genesis*, preserved by Eusebius, shows us that the author was familiar with the latest astronomical discoveries, especially the precession of the equinoxes. The writings in which the Fathers of the Church comment on the work of the six days of Creation, however, notably the commentaries of St. Basil and St. Ambrose, borrow but little from Hellenic physics; in fact, their tone would seem to indicate distrust in the teachings of Greek science, this distrust being engendered by two prejudices: In the first place, astronomy was becoming more and more the slave of astrology, the superstitions of which the Church diligently combated; in the second place, between the essential propositions of peripatetic physics and what we believe to be the teaching of Holy Writ, contradictions appeared; thus, Genesis was thought to teach the presence of water above the heaven of the fixed stars (the firmament), and this was incompatible with the Aristotelian theory concerning the natural place of the elements. The debates raised by this question gave St. Augustine an opportunity to establish wise exegetical rules, and he recommended that Christians not put forth lightly, as articles of faith, propositions contradicted by physical science based on careful experiments. St. Isidore of Seville (d. 636), a bishop, considered it legitimate for Christians to desire to know the teachings of profane science, and he labored to satisfy this curiosity. His *Etymologies* and *De natura rerum* are merely compilations of fragments borrowed from all the pagan and Christian authors with whom he was acquainted. At the height of the Latin Middle Ages, these works served as models for numerous encyclopedias, of which the *De natura rerum* by Bede (about 672–735) and the *De universo* by Rabanus Maurus (776–856) were the best known.

The sources from which the Christians of the West imbibed a knowledge of ancient physics, however, became daily more numerous, and to Pliny the Elder's *Natural History*, read by Bede, were added Chalcidius's commentary on Plato's *Timaeus* and Martianus Capella's *De nuptiis philologiae et mercurii*, these different works inspiring the physics of John Scotus Eriugena. Prior to A.D. 1000, a new Platonic work by Macrobius, a commentary on the *Somnium Scipionis*, was in great favor in the schools. Influenced by the various treatises already mentioned, Guillaume of Conches (1080–1150 or 1154) and the unknown author of *De mundi constitutione liber*—which, by the way, has been falsely attributed to Bede—set forth a planetary theory making Venus and Mercury satellites of the sun, but Eriugena went still further and made the sun also the center of the orbits of Mars and Jupiter. Had he but extended this hypothesis to Saturn, he would have merited the title of precursor of Tycho Brahe.

III. A Glance at Arabic Physics

The authors of whom we have heretofore spoken had only been acquainted with Greek science through the medium of Latin tradition, but the time came when it was to be much more completely revealed to the Christians of the West through the medium of Islamic tradition.

There is no Arabic science.[3] The wise men of Islam were always the more or less faithful disciples of the Greeks but were themselves devoid of all originality. For instance, they compiled many abridgments of Ptolemy's *Almagest*, made numerous observations, and constructed a great many astronomical tables, but they added nothing essential to the theories of astronomical motion; their only innovation in this respect—and, by the way, an unfortunate one—was the doctrine of the oscillatory motion of the equinoctial points, which the Middle Ages ascribed to Thabit ibn Qurrah (836–901) but which was probably the idea of al-Zarqali, who lived much later and made observations between 1060 and 1080. This motion was merely the adaptation of a mechanism conceived by Ptolemy for a totally different purpose.

In physics, Arabic scholars confined themselves to commentaries on the statements of Aristotle, their attitude being at times one of absolute servility. This intellectual servility to peripatetic teaching reached its climax in Abu'l ibn Rushd, whom Latin scholastics called Averroës (about 1120–1198) and who said: Aristotle "founded and completed logic, physics, and metaphysics . . . because none of those who have followed him up to our time, that is to say, for four hundred years, have been able to add anything to his writings or to detect therein an error of any importance." This unbounded respect for Aristotle's work impelled a great many Arabic philosophers to attack Ptolemy's *Astronomy* in the name of peripatetic

3. [See note 2.]

physics. The conflict between the hypotheses of eccentrics and epicycles was inaugurated by ibn Bajja, known to the scholastics as Avempace (d. 1138), and Abu Bakr ibn Tufayl, called Abubacer by the scholastics (d. 1185), and was vigorously conducted by Averroës, the protégé of Abubacer. Abu Ishaq al-Bitrugi al-Ishbili, known by the scholastics as Alpetragius, another disciple of Abubacer and a contemporary of Averroës, advanced a theory on planetary motion wherein he wished to account for the phenomena peculiar to the wandering stars by compounding rotations of homocentric spheres; his treatise, which was more neo-Platonic than peripatetic, seemed to be a Greek book altered or a simple plagiarism. Less inflexible in his peripateticism than Averroës and Alpetragius, Moses ben Maimon, called Maimonides (1139–1204), accepted Ptolemy's astronomy despite its incompatibility with Aristotelian physics, although he regarded Aristotle's sublunary physics as absolutely true.

IV. Arabic Tradition and Latin Scholasticism

It cannot be said exactly when the first translations of Arabic writings began to be received by the Christians of the West, but it was certainly previous to the time of Gerbert (Sylvester II, about 930–1003). Gerbert used treatises translated from the Arabic and containing instructions on the use of astronomical instruments, notably the astrolabe, to which instrument Hermann the Lame (1013–1054) devoted part of his researches. At the beginning of the twelfth century, the contributions of Islamic science and philosophy to Latin Christendom became more and more frequent and important. About 1120 or 1130, Adelard of Bath translated the *Elements* of Euclid and various astronomical treatises; in 1141, Peter the Venerable, Abbot of Cluny, found two translators, Hermann the Second (or the Dalmatian) and Robert of Retines, established in Spain. He engaged them to translate the Koran into Latin, and in 1143, these translators made Christendom acquainted with Ptolemy's planisphere. Under the direction of Raimond (Archbishop of Toledo, 1130; d. 1150), Domenicus Gundissalinus, Archdeacon of Segovia, began to collaborate with the converted Jew John of Luna, erroneously called John of Seville (Johannes Hispalensis). While John of Luna applied himself to works in mathematics, he also assisted Gundissalinus in translating into Latin a part of Aristotle's physics, the *De caelo* and the *Metaphysics*, besides treatises by Avicenna, al-Ghazali, al-Farabi, and perhaps Solomon ibn Gabirol (Avicebron). About 1134, John of Luna translated al-Farghani's treatise *Astronomy*, which was an abridgment of the *Almagest*, thereby introducing Christians to the

Ptolemaic system; at the same time, his translations, done in collaboration with Gundissalinus, familiarized Latins with the physical and metaphysical doctrines of Aristotle. Indeed, the influence of Aristotle's *Physics* was already apparent in the writings of the most celebrated masters of the school of Chartres (from 1121 until before 1155) and of Gilbert de la Porrée (1070–1154).

The abridgment of al-Farghani's *Astronomy*, translated by John of Luna, does not seem to have been the first work in which the Latins were able to read the exposition of Ptolemy's system; it was undoubtedly preceded by a more complete treatise, the *De scientia stellarum* of Albategnius (al-Battani), latinized by Plato of Tivoli about 1120. The *Almagest* itself was still unknown, however. Moved by a desire to read and translate Ptolemy's immortal work, Gerard of Cremona (d. 1187) left Italy and went to Toledo, eventually making the translation, which he finished in 1175. Besides the *Almagest*, Gerard rendered into Latin other works, of which we have a list comprising seventy-four different treatises. Some of these were writings of Greek origin and included a large portion of the works of Aristotle, a treatise by Archimedes, Euclid's *Elements* (completed by Hypsicles), and books by Hippocrates. Others were Arabic writings, such as the celebrated *Book of Three Brothers*, composed by the Banu Musa, *Optics* by ibn al-Haytham (the Alhazen of the Scholastics), *Astronomy* by Geber, and *De motu octavae sphaerae* by Thabit ibn Qurrah. Moreover, in order to spread the study of Ptolemaic astronomy, Gerard composed at Toledo his *Theoricae planetarum*, which, during the Middle Ages, became one of the classics of astronomical instruction. Beginners who obtained their first cosmographic information through the study of the *Sphaera*, written about 1230 by Joannes de Sacrobosco, could acquire a knowledge of eccentrics and epicycles by reading the *Theoricae planetarum* of Gerard of Cremona. In fact, until the sixteenth century, most astronomical treatises assumed the form of commentaries, on either the *Sphaera* or the *Theoricae planetarum*.

"Aristotle's philosophy," wrote Roger Bacon in 1267, "reached a great development among the Latins when Michael Scot appeared about 1230, bringing with him certain parts of the mathematical and physical treatises of Aristotle and his learned commentators." Among the Arabic writings made known to Christians by Michael Scot (before 1291; astrologer to Frederick II) were the treatises of Aristotle and the *Theory of Planets*, which Alpetragius had composed in accordance with the hypothesis of homocentric spheres. The translation of this last work was completed in 1217. By propagating among the Latins the commentaries on Averroës and on Alpetragius's theory of the planets, as well as a knowledge of the treatises of Aristotle, Michael Scot developed in them an intellectual disposition which might be termed Averroism and which consisted in a superstitious respect for the word of Aristotle and his commentator.

There was a metaphysical Averroism which, because professing the doctrine of the substantial unity of all human intellects, was in open conflict with Christian orthodoxy; but there was likewise a physical Averroism which, in its blind confidence in peripatetic physics, held as absolutely certain all that the latter taught on the subject of the celestial substance, rejecting in particular the system of epicycles and eccentrics in order to commend Alpetragius's astronomy of homocentric spheres.

Scientific Averroism found partisans even among those whose purity of faith constrained them to struggle against metaphysical Averroism and who were often peripatetics insofar as was possible without formally contradicting the teaching of the Church. For instance, William of Auvergne (d. 1249), who was the first to combat "Aristotle and his sectarians" on metaphysical grounds, was somewhat misled by Alpetragius's astronomy, which, moreover, he understood imperfectly. Albertus Magnus (1193 or 1205–1280) followed, to a great extent, the doctrine of Ptolemy, although he was sometimes influenced by the objections of Averroës or affected by Alpetragius's principles. Vincent of Beauvais, in his *Speculum quadruplex*, a vast encyclopaedic compilation published about 1250, seemed to attach great importance to the system of Alpetragius, borrowing the exposition of it from Albertus Magnus. Finally, even St. Thomas Aquinas (1227–1274) gave evidence of being extremely perplexed by the theory of eccentrics and epicycles, which justified celestial phenomena by contradicting the principles of peripatetic physics, and the theory of Alpetragius, which honored these principles but did not go so far as to represent their phenomena in detail.

This hesitation, so marked in the Dominican school, was hardly less remarkable in the Franciscan. Robert Grosseteste, or Greathead (1175–1253), whose influence on Franciscan studies was so great, followed the Ptolemaic system in his astronomical writings, his physics being imbued with Alpetragius's ideas. St. Bonaventure (1221–1274) wavered between doctrines which he did not thoroughly understand, and Roger Bacon (1214–1292), in several of his writings, weighed with great care the arguments that could be made to count for or against each of these two astronomical theories, without eventually making a choice. Bacon, however, was familiar with a method of figuration in the system of eccentrics and epicycles which Alhazen had derived from the Greeks; and in this figuration, all the motions acknowledged by Ptolemy were traced back to the rotation of solid orbs accurately fitted in one another. This representation, which refuted most of the objections raised by Averroës against Ptolemaic astronomy, contributed largely to propagating the knowledge of this astronomy, and it seems that the first of the Latins to adopt it and expatiate on its merits was the Franciscan Bernard of Verdun (end of thirteenth century), who had read Bacon's writings. In sublunary physics, the authors whom we have just mentioned did not show the hesitation that rendered astronomical doctrines so perplexing, but on almost all points, they adhered closely to peripatetic opinions.

V. The Science of Observation and Its Progress: Astronomers; The Statics of Jordanus; Theodoric of Freiberg; Pierre of Maricourt

Averroism had rendered scientific progress impossible, but fortunately in Latin Christendom it was to meet with two powerful enemies: the unhampered curiosity of human reason and the authority of the Church. Encouraged by the certainty resulting from experiments, astronomers rudely shook off the yoke which peripatetic physics had imposed on them. The School of Paris in particular was remarkable for its critical views and its freedom of attitude toward the argument of authority. In 1290, William of Saint-Cloud determined with wonderful accuracy the obliquity of the ecliptic and the time of the vernal equinox, and his observations led him to recognize the inaccuracies that marred the *Tables of Toledo*, drawn up by al-Zarkali. The theory of the precession of the equinoxes, conceived by the astronomers of Alfonso X of Castile, and the *Alphonsine Tables,* set up in accordance with this theory, gave rise in the first half of the fourteenth century to the observations, calculations, and critical discussions of Parisian astronomers, especially of John of Linières and his pupil John of Saxony or Connaught.

At the end of the thirteenth century and the beginning of the fourteenth, sublunary physics owed great advancement to the simultaneous efforts of geometers and experimenters—their method and discoveries being duly boasted of by Roger Bacon, who, however, took no important part in their labors. Jordanus de Nemore, a talented mathematician who, not later than about the beginning of the thirteenth century, wrote treatises on arithmetic and geometry, left a very short treatise on statics in which, side by side with erroneous propositions, we find the law of the equilibrium of the straight lever correctly established with the aid of the principle of virtual displacements. The treatise *De ponderibus* by Jordanus provoked research on the part of various commentators, and one of these, whose name is unknown and who must have written before the end of the thirteenth century, drew, from the same principle of virtual displacements, demonstrations, admirable in exactness and elegance, of the law of the equilibrium of the bent lever and of the apparent weight (*gravitas secundum situm*) of a body on an inclined plane.

Alhazen's *Treatise on Perspective* was read thoroughly by Roger Bacon and his contemporaries, John Peckham (1228–1291), the English Franciscan; he gave a summary of it. About 1270, Witelo (or Vitellio) composed an exhaustive ten-volume treatise on optics, which remained a classic until the time of Kepler, who wrote a commentary on it.

Albertus Magnus, Roger Bacon, John Peckham, and Witelo were deeply interested in the theory of the rainbow, and, like the ancient meteorologists, they all took the rainbow to be the image of the sun reflected in a sort of concave mirror formed by a cloud resolved into rain. In 1300, Theodoric of Freiberg proved by means of carefully conducted experiments, in which he used glass balls filled with water, that the rays which render the bow visible have been reflected on the inside of the spherical drops of water, and he traced with great accuracy the course of the rays which produce the rainbows respectively.

The system of Theodoric of Freiberg—at least that part relating to the primary rainbow—was reproduced about 1360 by Themon, "Son of the Jew" (Themo Judaeus); from his commentary on *Meteors*, it passed down to the days of the Renaissance when, having been somewhat distorted, it reappeared in the writings of Alessandro Piccolomini, Simon Porta, and Marco and Antonio de Dominis; thus, it was propagated until the time of Descartes.

The study of the magnet had also made great progress in the course of the thirteenth century; the permanent magnetization of iron, the properties of the magnetic poles, and the direction of the Earth's action exerted on these poles or of their action on one another are all found very accurately described in a treatise written in 1269 by Pierre of Maricourt (Petrus Peregrinus). Like the work of Theodoric of Freiberg on the rainbow, the *Epistola de magnete* by Maricourt was a model of the art of logical sequence between experiment and deduction.

VI. The Articles of Paris (1277): The Possibility of Vacuum

The University of Paris was uneasy because of the antagonism between Christian dogmas and certain peripatetic doctrines, and on several occasions it combated Aristotelian influence. In 1277, Etienne Tempier,

Bishop of Paris, acting on the advice of the theologians of the Sorbonne, condemned a great number of errors, some of which emanated from the astrology and others from the philosophy of the peripatetics. Among these errors considered dangerous to faith were several which might have impeded the progress of physical science, and hence it was that the theologians of Paris declared erroneous the opinion maintaining that God Himself could not give the entire universe a rectilinear motion, as the universe would then leave a vacuum behind it, and also declared false the notion that God could not create several worlds. These condemnations destroyed certain essential foundations of peripatetic physics because although in Aristotle's system such propositions were ridiculously untenable, belief in Divine Omnipotence sanctioned them as possible while waiting for science to confirm them as true. For instance, Aristotle's physics treated the existence of an empty space as a pure absurdity; in virtue of the "Articles of Paris," Richard of Middleton (about 1280) and, after him, many masters in Paris and Oxford admitted that the laws of nature are certainly opposed to the production of empty space but that the realization of such a space is not, in itself, contrary to reason; thus, without absurdity, one could argue on vacuum and on motion in a vacuum. Next, in order that such arguments might be legitimated, it was necessary to create that branch of mechanical science known as dynamics.

VII. The Earth's Motion: Oresme

The "Articles of Paris" about equally supported the question of the Earth's motion and furthered the progress of dynamics by regarding vacuum as something conceivable.

Aristotle maintained that the first heaven (the firmament) moved with a uniform rotary motion and that the Earth was absolutely stationary, and because these two propositions necessarily resulted from the first principles relative to time and place, it would have been absurd to deny them. By declaring that God could endow the World with a rectilinear motion, however, the theologians of the Sorbonne acknowledged that these two Aristotelian propositions could not be imposed as a logical necessity; thenceforth, while continuing to admit that, as a fact, the Earth was immovable and that the heavens moved with a rotary diurnal motion, Richard of Middleton and Duns Scotus (about 1275–1308) began to formulate hypotheses to the effect that these bodies were animated by other motions, and the entire school of Paris adopted the same opinion. Soon, however, the Earth's motion was taught in the School of Paris not as a possibility but as a reality. In fact, in the specific setting forth of certain information given by Aristotle and Simplicius, a principle was formulated which for three centuries was to play a great role in statics: That every heavy body tends to unite its center of gravity with the center of the Earth.

When writing his *Questions* on Aristotle's *De caelo* in 1368, Albert of Helmstadt (or of Saxony) admitted this principle, which he applied to the entire mass of the terrestrial element. The center of gravity of this mass is constantly inclined to place itself in the center of the universe, but, within the terrestrial mass, the position of the center of gravity is constantly changing. The principal cause of this variation is the erosion brought about by the streams and rivers that continually wear away the land surface, deepening its valleys and carrying off all loose matter to the bed of the sea, thereby producing a displacement of weight which entails a ceaseless change in the position of the center of gravity. Now, in order to replace this center of gravity at the center of the universe, the Earth moves without ceasing; meanwhile, a slow but perpetual exchange is being effected between the continents and the oceans. Albert of Saxony ventured so far as to think that these small and incessant motions of the Earth could explain the phenomena of the precession of the equinoxes. The same author declared that one of his masters, whose name he did not disclose, announced himself in favor of the daily rotation of the Earth, inasmuch as he refuted the arguments that were opposed to this motion. This anonymous master had a thoroughly convinced disciple in Nicole Oresme, who, in 1377, being then Canon of Rouen and later Bishop of Lisieux, wrote a French commentary on Aristotle's treatise *De caelo*, maintaining with as much force as clarity that neither experiment nor argument could determine whether the daily motion belonged to the firmament of the fixed stars or to the Earth. He also showed how to interpret the difficulties encountered in "the Sacred Scriptures wherein it is stated that the sun turns, etc. It might be supposed that here Holy Writ adapts itself to the common mode of human speech, as also in several places, for instance, where it is written that God repented Himself, and was angry and calmed Himself and so on, all of which is, however, not to be taken in a strictly literal sense." Finally, Oresme offered several considerations favorable to the hypothesis of the Earth's daily motion. In order to refute one of the objections raised by the peripatetics against this point, Oresme was led to explain how, in spite of this motion, heavy bodies seemed to fall in a vertical line; he admitted their real motion to be composed of a fall in a vertical line and a diurnal rotation identical with that which they would have if bound to the Earth. This is precisely the principle to which Galileo was afterward to turn.

VIII. Plurality of Worlds

Aristotle maintained the simultaneous existence of several worlds to be an absurdity, his principal argument being drawn from his theory of gravity, whence he concluded that two distinct worlds could not coexist and be each surrounded by its elements; therefore it would be ridiculous to compare each of the planets to an earth similar to ours. In 1277, the theologians of Paris condemned this doctrine as a denial of the creative omnipotence of God; Richard of Middletown and Henry of Ghent (who wrote in about 1280), Guillaume Varon (who wrote a commentary on the *Sentences* about 1300), and, about 1320, Jean de Bassols, William of Ockham (d. after 1347), and Walter Burley (d. about 1343) did not hesitate to declare that God could create other worlds similar to ours. This doctrine, adopted by several Parisian masters, required that the theory of gravity and natural place developed by Aristotle be thoroughly changed; in fact, the following theory was substituted for it. If some part of the elements forming a world were to be detached from it and driven far away, its tendency would be to move toward the world to which it belongs and from which it was separated; the elements of each world are inclined to arrange themselves so that the heaviest will be in the center and the lightest on the surface. This theory of gravity appeared in the writings of Jean Buridan of Béthune, who became rector of the University of Paris in 1327, teaching at that institution until about 1360; and in 1377, the same theory was formally proposed by Oresme. It was also destined to be adopted by Copernicus and his first followers and to be maintained by Galileo, William Gilbert, and Otto von Guericke.

IX. Dynamics: Theory of Impetus; Inertia; Celestial and Sublunary Mechanics Identical

If the School of Paris completely transformed the peripatetic theory of gravity, it was equally responsible for the overthrow of Aristotelian dynamics. Convinced that in all motion the mover should be directly contiguous to the body moved, Aristotle had proposed a strange theory of the motion of projectiles. He held that the projectile was moved by the fluid medium, whether air or water, through which it passed and this by virtue of the vibration brought about in the fluid at the moment of throwing and spread through it. In the sixth century of our era, this explanation was strenuously opposed by the Christian Stoic Joannes Philoponus, according to whom the projectile was moved by a certain power communicated to it at the instant of throwing; however, despite the objections raised by Philoponus, Aristotle's various commentators, particularly Averroës, continued to attribute the motion of the projectile to the disturbance of the air, and Albertus Magnus, St. Thomas Aquinas, Roger Bacon, Giles of Rome, and Walter Burley persevered in maintaining this error. By means of most spirited argumentation, William of Ockham made known the complete absurdity of the peripatetic theory of the motion of projectiles. Going back to Philoponus's thesis, Buridan gave the name *impetus* to the virtue or power communicated to the projectile by the hand or instrument throwing it; he declared that in any given body in motion, this impetus was proportional to the velocity and that, in different bodies in motion propelled by the same velocity, the quantities of impetus were proportional to the mass or quantity of matter, defined as it was afterwards defined by Newton.

In a projectile, impetus is gradually destroyed by the resistance of air or some other medium and is also destroyed by the natural gravity of the body in motion, which gravity is opposed to the impetus if the projectile is thrown upward; this struggle explains the different peculiarities of the motion of projectiles. In a falling body, gravity comes to the assistance of impetus, which it increases at every instant; hence, the velocity of the fall is constantly increasing.

With the assistance of these principles concerning impetus, Buridan accounts for the swinging of the pendulum. He likewise analyzes the mechanism of impact and rebound and, in this connection, puts forth correct views on the deformations and elastic reactions that arise in the contiguous parts of two bodies coming into collision. Nearly all this doctrine of impetus is transformed into a correct mechanical theory if one is careful to substitute the expression *vis viva* for *impetus*. The dynamics expounded by Buridan were adopted in their entirety by Albert of Saxony, Oresme, Marsilius of Inghen, and the entire School of Paris. Albert of Saxony appended thereto the statement that the velocity of a falling body must be proportional either to the time elapsed from the beginning of the fall or to the distance traversed during this time. In a projectile, the impetus is gradually destroyed either by the resistance of the medium or by the contrary tendency of the gravity natural to the body. When these causes of destruction do not exist, the impetus remains perpetually the same, as in the case of a millstone exactly centered and not rubbing on its axis; once set in motion, it will turn indefinitely with the same swiftness. It was in this form that the law of inertia first became evident to Buridan and Albert of Saxony.

The conditions manifested in this hypothetical millstone are realized in the celestial orbs, as in these, neither friction nor gravity impedes motion; hence, it may be admitted that each celestial orb moves indefinitely by virtue of a suitable impetus communicated to it by God at the moment of creation. It is useless to imitate Aristotle and his commentators by attributing the motion of each orb to a presiding spirit. This was the opinion proposed by Buridan and adopted by Albert of Saxony; and while formulating a doctrine from which modern dynamics was to spring, these masters understood that the same dynamics governs both celestial and sublunary bodies. Such an idea was directly opposed to the essential distinction established by ancient physics between these two kinds of bodies. Moreover, following William of Ockham, the masters of Paris rejected this distinction; they acknowledged that the matter constituting celestial bodies was of the same nature as that constituting sublunary bodies and that, if the former remained perpetually the same, it was not because they were, by nature, incapable of change and destruction, but simply because the place in which they were located contained no agent capable of corrupting them. A century elapsed between the condemnations pronounced by Etienne Tempier (1277) and the editing of the *Traité du ciel et du monde* by Oresme (1377), and within that time, all the essential principles of Aristotle's physics were undermined and the great controlling ideas of modern science formulated. This revolution was mainly the work of Oxford Franciscans such as Richard of Middletown, Duns Scotus, and William of Ockham and of masters in the School of Paris, heirs to the tradition inaugurated by these Franciscans; among the Parisian masters, Buridan, Albert of Saxony, and Oresme were in the foremost rank.

X. Propagation of the Doctrines of the School of Paris in Germany and Italy; Peurbach and Regiomontanus; Nicholas of Cusa; da Vinci

The great Western schism involved the University of Paris in politico-religious quarrels of extreme violence; the misfortunes brought about by the conflict between the Armagnacs and Burgundians and by the Hundred Years' War completed what these quarrels had begun, and the wonderful progress made by science during the fourteenth century in the University of Paris suddenly ceased. The schism, however, contributed to the diffusion of Parisian doctrines by driving out of Paris a large number of brilliant men who had taught there with marked success. In 1386, Marsilius of Inghen (d. 1396), who had been one of the most gifted professors of the University of Paris, became rector of the infant University of Heidelberg, where he introduced the dynamic theories of Buridan and Albert of Saxony.

About the same time, another master, reputedly of Paris, Heinrich Heimbuch of Langenstein (or of Hesse), was chiefly instrumental in founding the University of Vienna and, besides his theological knowledge, brought there the astronomical tradition of John of Linières and John of Saxony. This tradition was carefully preserved in Vienna, was magnificently developed there throughout the fifteenth century, and paved the way for Georg Peurbach (1423–1461) and his disciple Johann Müller of Königsberg, surnamed Regiomontanus (1436–1476). It was to the writing of theories calculated to make the Ptolemaic system known, to the designing and constructing the exact instruments, to the multiplying of observations, and to the preparing of tables and almanacs (ephemerides) more accurate than those used by astronomers up to that time that Peurbach and Regiomontanus devoted their prodigious energy. By perfecting all the details of Ptolemy's theories, which they never called into question, they were most helpful in bringing to light the defects of these theories and in preparing the materials by means of which Copernicus was to build his new astronomy.

Averroism flourished in the Italian universities of Padua and Bologna, which were noted for their adherence to peripatetic doctrines. Still, from the beginning of the fifteenth century, the opinions of the School of Paris began to find their way into these institutions, thanks to the teaching of Paolo Nicoletti of Venice (flourished about 1420). It was there developed by his pupil Gaetan of Tiene (d. 1465). These masters devoted special attention to propagating the dynamics of impetus in Italy.

About the time that Paul of Venice was teaching at Padua, Nicholas of Cusa came there to take his doctorate in law. Whether it was then that the latter became initiated in the physics of the School of Paris matters little, because in any event it was from Parisian physics that he adopted those doctrines that smacked least of peripateticism. He became thoroughly conversant with the dynamics of impetus and, like Buridan and Albert of Saxony, attributed the motion of the celestial spheres to the impetus which God had communicated to them in creating them and which was perpetuated because, in these spheres, there was no element of destruction. He admitted that the Earth moved incessantly and that its motion might be the cause of the precession of the equinoxes. In a note discovered long after his death, he went so far as to attribute to the Earth a daily rotation. He imagined that the sun, the moon, and the planets were so many systems, each of which contained an earth and elements analogous to our Earth and elements, and to account for the action of gravity in each of these systems, he closely followed the theory of gravity advanced by Oresme.

Leonardo da Vinci (1452–1519) was perhaps more thoroughly convinced of the merits of the Parisian physics than any other Italian master. A keen observer and endowed with insatiable curiosity, he had studied a great number of works, among which we may mention the various treatises of the School of Jordanus, various books by Albert of Saxony, and in all likelihood the works of Nicholas of Cusa; then, profiting by the learning of these scholars, he formally enunciated or simply intimated many new ideas. The statics of the School of Jordanus led him to discover the law of the composition of concurrent forces, stated as follows: The two component forces have equal moments as regards the direction of the resultant, and the resultant and one of the components have equal moments as regards the direction of the other component. The statics derived from the properties which Albert of Saxony attributed to the center of gravity caused da Vinci to recognize the law of the polygon of support and to determine the center of gravity of a tetrahedron. He also presented the law of the equilibrium of two liquids of different density in communicating tubes, and the principle of virtual displacements seems to have occasioned his acknowledgment of the hydrostatic law known as Pascal's. Da Vinci continued to meditate on the properties of impetus, which he called *impeto* or *forza*, and the propositions that he formulated on the subject of this power often showed a fairly clear discernment of the law of the conservation of energy. These propositions led him to remarkably correct and accurate conclusions concerning the impossibility of perpetual motion. Unfortunately, he misunderstood the pregnant explanation, afforded by the theory of impetus, regarding the acceleration of falling bodies, and, like the peripatetics, he attributed this acceleration to the impulsion of the encompassing air. By way of compensation, however, he distinctly asserted that the velocity of a body that falls freely is proportional to the time occupied in the fall, and he understood in what way this law extends to a fall on an inclined plane. When he wished to determine how the path traversed by a falling body is connected with the time occupied in the fall, he was confronted by a difficulty which, in the seventeenth century, was likewise to baffle Baliani and Gassendi.

Da Vinci was much engrossed in the analysis of the deformations and elastic reactions which cause a body to rebound after it has struck another, and this doctrine, formulated by Buridan, Albert of Saxony, and Marsilius of Inghem, he applied in such a way as to draw from it the explanation of the flight of birds. This flight is an alternation of falls during which the bird compresses the air beneath it, and of rebounds due to the elastic force of this air. Until the great painter discovered this explanation, the question of the flight of birds was always looked on as a problem in statics and was likened to the swimming of fish in water. Da Vinci attached great importance to the views developed by Albert of Saxony in regard to the Earth's equilibrium. Like the Parisian master, he held that the center of gravity within the terrestrial mass is constantly changing under the influence of erosion and that the Earth is continually moving so as to bring this center of gravity to the center of the World. These small, incessant motions eventually bring to the surface of the continents those portions of Earth that once occupied the bed of the ocean, and to place this assertion of Albert of Saxony beyond the range of doubt, da Vinci devoted himself to the study of fossils and to extremely cautious observations which made him the creator of Stratigraphy. In many passages in his notes, da Vinci asserts, like Nicholas of Cusa, that the moon and the other wandering stars are worlds analogous to ours, that they carry seas upon their surfaces and are surrounded by air; and the development of this opinion led him to talk of the gravity binding to each of these stars the elements that belonged to it. On the subject of this gravity he professed a theory similar to Oresme's. Hence it would seem that, in almost every particular, da Vinci was a faithful disciple of the great Parisian masters of the fourteenth century—of Buridan, Albert of Saxony, and Oresme.

XI. Italian Averroism and Its Tendencies to Become Routine; Attempts at Restoring the Astronomy of Homocentric Spheres

While, through the anti-peripatetic influence of the School of Paris, da Vinci reaped a rich harvest of discoveries, innumerable Italians devoted themselves to the sterile worship of defunct ideas with a servility that was truly astonishing. The Averroists did not wish to acknowledge as true anything out of conformity with the ideas of Aristotle as interpreted by Averroës; with Pompanazzi (1462–1526), the Alexandrists, seeking their inspiration further in the past, refused to understand Aristotle in a manner other than that in which he had been understood by Alexander of Aphrodisias; and the humanists, solicitous only for purity of form, would not consent to use any technical language whatsoever and rejected all ideas that were not sufficiently vague to be attractive to orators and poets. Thus, Averroists, Alexandrists, and humanists proclaimed a truce to their vehement discussions so as to combine against the "language of Paris," the "logic of Paris," and the "physics of Paris." It is difficult to conceive the absurdities to which these minds were led by their slavish surrender to routine. A great number of physicists, rejecting the Parisian theory of impetus, returned to the untenable dynamics of Aristotle and maintained that the projectile was moved by the ambient air. In 1409, Nicoletto Vernia of Chieti, an Averroist professor at Padua, taught that if a heavy body fell, it was in consequence of the motion of the air surrounding it.

A servile adoration of peripateticism prompted many so-called philosophers to reject the Ptolemaic system, the only one which, at that time, could satisfy the legitimate exigencies of astronomers, and to readopt the hypothesis of homocentric spheres. They held as null and void the innumerable observations that showed changes in the distance of each planet from the Earth. Alessandro Achillini of Bologna (1463–1512), an uncompromising Averroist and a strong opponent of the theory of impetus and of all Parisian doctrines, inaugurated, in his treatise *De orbibus* (1498), a strange reaction against Ptolemaic astronomy; Agostino Nifo (1473–1538) labored for the same end in a work that has not come down to us; Girolamo Fracastoro (1483–1553) gave us, in 1535, his book *Homocentricorum*; and Giovanni Battista Amico (1536) and Giovanni Antonio Delfino (1559) published small works in an endeavor to restore the system of homocentric spheres.

XII. The Copernican Revolution

Although directed by tendencies diametrically opposed to the true scientific spirit, the efforts made by Averroists to restore the astronomy of homocentric spheres were perhaps a stimulus to the progress of science, inasmuch as they accustomed physicists to the thought that the Ptolemaic system was not the only astronomical doctrine possible, or even the best that could be desired. Thus, in their own way, the Averroists paved the way for the Copernican revolution. The movements forecasting this revolution were noticeable in the middle of the fourteenth century in the writings of Nicholas of Cusa and in the beginning of the fifteenth century in the notes of da Vinci, both of these eminent scientists being well versed in Parisian physics.

Celio Calcagnini proposed, in his turn, to explain the daily motion of the stars by attributing to the Earth a rotation from west to east, complete in one sidereal day. His dissertation, *Quod coelum stet, terra vero moveatur*, although apparently written about 1530, was not published until 1544, when it appeared in a posthumous edition of the author's works. Calcagnini declared that the Earth, originally in equilibrium in the center of the universe, received a first impulse which imparted to it a rotary motion, and this motion, to which nothing was opposed, was indefinitely preserved by virtue of the principle set forth by Buridan and accepted by Albert of Saxony and Nicholas of Cusa. According to Calcagnini, the daily rotation of the Earth was accompanied by the oscillation which explained the movement of the precession of the equinoxes. Another oscillation set the waters of the sea in motion and determined the ebb and flow of the tides. This last hypothesis was to be maintained by Andrea Cesalpino (1519–1603) in his *Quaestiones peripateticae* (1569) and to inspire Galileo, who, unfortunately, was to seek in the phenomena of the tides his favorite proof of the Earth's rotation.

The *De revolutionibus orbium caelestium libri sex* were printed in 1543, a few months after the death of Copernicus (1473–1543), but the principles of the astronomical system proposed by this man of genius had been published as early as 1539 in the *Narratio prima* of his disciple, Joachim Rheticus (1514–1574). Copernicus adhered to the ancient astronomical hypotheses, which claimed that the World was spherical and limited and that all celestial motions were decomposable into circular and uniform motions; but he held that the firmament of fixed stars was immovable, as also was the sun, which was placed in the center of this firmament. To the Earth he attributed three motions: a circular motion by which the center of the Earth described with uniform velocity a circle situated in the plane of the ecliptic and eccentric to the sun; a daily rotation on an axis inclined toward the ecliptic; and finally, a rotation of this axis around an axis normal to the ecliptic and passing through the center of the Earth. The time occupied by this last rotation was a little longer than that required for the circular motion of the center of the Earth, which produced the phenomenon of the precession of the equinoxes. To the five planets Copernicus ascribed motions analogous to those with which the Earth was provided, and he maintained that the moon moved in a circle around the Earth.

Of the Copernican hypotheses, the newest was that according to which the Earth moved in a circle around the sun. From the days of Aristarchus of Samos and Seleucus, no one had adopted this view. Medieval astronomers had all rejected it because they supposed that the stars were much too close to the Earth and the sun, and that an annual circular motion of the Earth might give the stars a perceptible parallax. Still, on the other hand, we have seen that various authors had proposed to attribute to the Earth one or the other of the two motions which Copernicus added to the annual motion. To defend the hypothesis of the daily motion of the Earth against the objections formulated by peripatetic physics, Copernicus invoked exactly the same reasons as Oresme, and in order to explain how each planet retains the various parts of its elements, he adopted the theory of gravity proposed by the eminent master. Copernicus showed himself to be an adherent of Parisian physics even in the following opinion, enunciated accidentally: The acceleration of the fall of heavy bodies is explained by the continual increase which impetus receives from gravity.

XIII. Fortunes of the Copernican System in the Sixteenth Century

Copernicus and his disciple Rheticus probably regarded the motions which their theory ascribed to the Earth and the planets, the sun's rest and that of the firmament of fixed stars, as the real motions or real rest of these bodies. The *De revolutionibus orbium caelestium libri sex* appeared with an anonymous preface which inspired an entirely different idea. This preface was the work of the Lutheran theologian Osiander (1498–1552), who therein expressed the opinion that the hypotheses proposed by philosophers in general, and by Copernicus in particular, were in no way calculated to acquaint us with the reality of things: It is neither necessary that these hypotheses be true, nor even that they be likely, but one thing is sufficient; namely, that the calculation to which they lead agrees with observation.[4] Osiander's view of astronomical hypotheses was not new. Even in the days of Grecian antiquity, a number of thinkers had maintained that the sole object of these hypotheses was to "save appearances, *sozein ta phainomena*"; and in the Middle Ages, as well as in antiquity, this method continued to be that of philosophers who wished to make use of Ptolemaic astronomy while at the same time upholding the peripatetic physics absolutely incompatible with this astronomy. Osiander's doctrine was therefore readily received, first by astronomers who, without believing the Earth's motion to be a reality, accepted and admired the kinetic combinations conceived by Copernicus, as these combinations provided them with better means than could be offered by the Ptolemaic system for figuring out the motion of the moon and the phenomena of the precession of the equinoxes.

4. *Neque enim necesse est eas hypotheses esse veras, imo, ne verisimiles quidem, sed sufficit hoc unum si calculum observationibus congruentem exhibeant.*

One of the astronomers who most distinctly assumed this attitude in regard to Ptolemy's system was Erasmus Reinhold (1511–1553), who, though not admitting the Earth's motion, professed a great admiration for the system of Copernicus and used it in computing new astronomical tables, the *Prutenicae tabulae* (1551), which were largely instrumental in introducing to astronomers the kinetic combinations originated by Copernicus. The *Prutenicae tabulae* were especially employed by the commission which, in 1582, effected the Gregorian reform of the calendar. While not believing in the Earth's motion, the members of this commission did not hesitate to use tables founded on a theory of the precession of the equinoxes and attributing a certain motion to the Earth.

The freedom permitting astronomers to use all hypotheses qualified to account for phenomena, however, was soon restricted by the exigencies of peripatetic philosophers and Protestant theologians. Osiander had written his celebrated preface to Copernicus's book with a view toward warding off the attacks of theologians, but in this he did not succeed. Martin Luther, in his *Tischrede*, was the first to express indignation at the impiety of those who admitted the hypothesis of solar rest. Melanchthon, though acknowledging the purely astronomical advantages of the Copernican system, strongly combated the hypothesis of the Earth's motion (1549), not only with the aid of arguments furnished by peripatetic physics but also, and chiefly, with the assistance of numerous texts taken from Holy Writ. Kaspar Peucer (1525–1602), Melanchthon's son-in-law, while endeavoring to have his theory of the planets harmonize with the progress which the Copernican system had made in this regard, nevertheless rejected the Copernican hypotheses as absurd (1571).

It then came to be demanded of astronomical hypotheses that not only, as Osiander had desired, must the results of their calculations be conformable to facts but also that they must not be refutable "either in the name of the principles of physics or in the name of the authority of the Sacred Scriptures." These criteria were explicitly formulated in 1578 by a Lutheran, the Danish astronomer Tycho Brahe (1546–1601), and it was precisely by virtue of these two requirements that the doctrines of Galileo were condemned by the Inquisition in 1616 and 1633. Eager not to admit any hypothesis that would conflict with Aristotelian physics or be contrary to the letter of the Sacred Scriptures, and yet most desirous to retain all the astronomical advantages of the Copernican system, Tycho Brahe proposed a new system which virtually consisted in leaving the Earth motionless and in moving the other heavenly bodies in such a way that their displacement with regard to the Earth might remain the same as in the system of Copernicus. Moreover, though posing as the defender of Aristotelian physics, Tycho Brahe dealt it a disastrous blow. In 1572, a star, until then unknown, appeared in the constellation of Cassiopeia, and in showing accurate observations that the new astral body was really a fixed star, Tycho Brahe proved conclusively that the celestial world was not, as Aristotle would have had us believe, formed of a substance exempt from generation and destruction.

The Church had not remained indifferent to the hypothesis of the Earth's motion until the time of Tycho Brahe, as it was among her members that this hypothesis had found its first defenders, counting adherents even in the extremely orthodox University of Paris. At the time of defending this hypothesis, Oresme was Canon of Rouen, and immediately afterward he was promoted to the Bishopric of Lisieux; Nicholas of Cusa was Bishop of Brixen and cardinal and was entrusted with important negotiations by Eugenius IV, Nicholas V, and Pius II; Calcagnini was prothonotary Apostolic; Copernicus was Canon of Thorn, and it was Cardinal Schomberg who urged him to publish his work, the dedication of which was accepted by Paul III. Besides, Oresme had made clear how to interpret the scriptural passages claimed to be opposed to the Copernican system, and in 1584, Didacus a Stunica of Salamanca found in Holy Writ texts which could be invoked with as much certainty in favor of the Earth's motion. In 1595, however, the Protestant senate of the University of Tübingen compelled Kepler to retract the chapter in his *Mysterium cosmographicum* in which he had endeavored to make the Copernican system agree with Scripture.

Christopher Clavius (1537–1612), a Jesuit and one of the influential members of the commission that reformed the Gregorian calendar, seemed to be the first Catholic astronomer to adopt the double test imposed on astronomical hypotheses by Tycho Brahe and to decide (1581) that the suppositions of Copernicus were to be rejected, as opposed both to peripatetic physics and to Scripture. On the other hand, at the end of his life, and under the influence of Galileo's discoveries, Clavius appeared to have assumed a far more favorable attitude toward Copernican doctrines. The enemies of Aristotelian philosophy gladly adopted the system of Copernicus, considering its hypotheses as so many propositions physically true—this being the case with Pierre de La Rameé, called Petrus Ramus (1502–1572) and especially with Giordano Bruno (about 1550–1600). The physics developed by Bruno, in which he incorporated the Copernican hypothesis, proceeded from Nicole Oresme and Nicholas of Cusa, but chiefly from the physics taught in the University of Paris in the fourteenth century. The infinite extent of the universe and the plurality of worlds were admitted as possible by many theologians at the end of the thirteenth century, and the theory of the slow motion which gradually causes the central portions of the Earth to work to the surface had been taught by Albert of Saxony before it attracted the attention of da Vinci. The solution of peripatetic arguments against the Earth's motion and the theory of gravity called forth by the comparison of the planets with the Earth would appear to have been borrowed by Bruno from Oresme. The apostasy and heresies for which Bruno was condemned in 1600 had nothing to do with the physical doctrines he had espoused, which included in particular Copernican astronomy. In fact, it does not seem that in the sixteenth century, the Church manifested the slightest anxiety concerning the system of Copernicus.

XIV. Theory of the Tides

It is undoubtedly to the great voyages that shed additional luster on the close of the fifteenth century that we must attribute the importance assumed in the sixteenth century by the problem of the tides and the great progress made at that time toward the solution of this problem. The correlation existing between the phenomenon of high and low tide and the course of the moon was known even in ancient times. Posidonius accurately described it; the Arabic astronomers were also familiar with it, and the explanation given of it in the ninth century by Albumazar in his *Introductorium magnum ad astronomiam* remained a classic throughout the Middle Ages. The observation of tidal phenomena naturally led to the supposition that the moon attracted the waters of the ocean, and in the thirteenth century, William of Auvergne compared this attraction with that of the magnet for iron. The mere attraction of the moon, however, did not suffice to account for the alternation of spring and neap tides, which phenomenon clearly indicated a certain intervention of the sun. In his *Questions sur les livres des meteores*, which appeared during the latter half of the fourteenth century, Themo Judaeus introduced, in a vague sort of way, the idea of superposing two tides, the one due to the sun and the other to the moon.

In 1528, this idea was clearly endorsed by Federico Grisogono of Zara, a Dalmatian who taught medicine at Padua. Grisogono declared that under the action of the moon exclusively, the sea would assume an ovoid shape, its major axis being directed toward the center of the moon; that the action of the sun would also give it an ovoid shape, less elongated than the first, its major axis being directed toward the center of the sun; and that the variation of sea level, at all times and in all places, was obtained by adding the elevation or depression produced by the solar tide to the elevation or depression produced by the lunar tide. In 1557, Girolamo Cardano accepted and briefly explained Grisogono's theory. In 1559, a posthumous work by Delfino gave a description of the phenomena of the tides, identical with that deduced from the mechanism conceived by Grisogono. The doctrine of the Dalmatian physician was reproduced by that of Paolo Gallucci in 1588 and by that of Annibale Raimondo in 1589; and in 1600, Claude Duret, who had plagiarized Delfino's treatise, published in France the description of the tides given in that work.

XV. Statics in the Sixteenth Century: Stevin

When writing on statics, Cardano drew on two sources: the writings of Archimedes and the treatises of the School of Jordanus. In addition, he probably plagiarized the notes left by da Vinci, and it was perhaps from this source that he took the theorem: A system endowed with weight is in equilibrium when the center of gravity of this system is the lowest possible.

Nicolo Tartaglia (about 1500–1557), Cardano's antagonist, shamelessly purloined a supposedly forgotten treatise by one of Jordanus's commentators. Ferrari, Cardano's faithful disciple, harshly rebuked Tartaglia for the theft, which nevertheless had the merit of reestablishing the vogue of certain discoveries of the thirteenth century, especially the law of the equilibrium of a body supported by an inclined plane. By another and no less barefaced plagiarism, Tartaglia published under his own name a translation of Archimedes's *Treatise on Floating Bodies* made by William of Moerbeke at the end of the thirteenth century. This publication, dishonest though it was, helped to give prominence to the study of Archimedes's mechanical labors, which study exerted the greatest influence on the progress of science at the end of the sixteenth century—the blending of Archimedean mathematics with Parisian physics, generating the movement that terminated in Galileo's work. The translation and explanation of the works of Archimedes enlisted the attention of geometers such as Francesco Maurolico of Messina (1494–1575) and Federico Commandino of Urbino (1509–1575), and these two authors, continuing the work of the great Syracusan, determined the position of the center of gravity of various solids. In addition, Commandino translated and explained Pappus's mathematical *Collection* and the fragment of *Mechanics* by Heron of Alexandria appended thereto. Admiration for these monuments of ancient science

inspired a number of Italians with a profound contempt for medieval statics. The fecundity of the principle of virtual displacements, so happily
employed by the School of Jordanus, was ignored; and, deprived of the
laws discovered by this school and of the additions made to them by da
Vinci, the treatises on statics written by overenthusiastic admirers of the
Archimedean method were notably deficient. Among the authors of these
treatises, Guidobaldo dal Monte (1545–1607) and Giovanni Battista Benedetti (1530–1590) deserve special mention.

Of the mathematicians who, in statics, claimed to follow exclusively
the rigorous methods of Archimedes and the Greek geometers, the most
illustrious was Simon Stevin of Bruges (1548–1620). Through him, the
statics of solid bodies recovered all that had been gained by the School of
Jordanus and da Vinci and lost by the contempt of such men as
Guidobaldo del Monte and Benedetti. The law of the equilibrium of the
lever, one of the fundamental propositions of which Stevin made use, was
established by him with the aid of an ingenious demonstration which
Galileo was also to employ, and which is found in a small anonymous work
of the thirteenth century. In order to confirm another essential principle
of his theory, the law of the equilibrium of a body on an inclined plane,
Stevin resorted to the impossibility of perpetual motion, which had been
affirmed with great precision by da Vinci and Cardano. Stevin's chief
glory lay in his discoveries in hydrostatics; and the determination of the
extent and point of application of the pressure on the slanting inner side
of a vessel by the liquid contained therein was in itself sufficient to entitle
this geometer from Bruges to a foremost place among the creators of the
theory of the equilibrium of fluids. Benedetti was on the point of enunci-
ating the principle known as Pascal's law, and an insignificant addition
permitted Mersenne to infer this principle and the idea of the hydraulic
press from what the Italian geometer had written. Benedetti had justified
his propositions by using as an axiom the law of the equilibrium of liquids
in communicating vessels; prior to this time, da Vinci had followed the
same logical proceeding.

XVI. Dynamics in the Sixteenth Century

The geometers who, in spite of the stereotyped methods of Averroism and
the banter of humanism, continued to cultivate the Parisian dynamics of
impetus were rewarded by splendid discoveries. Dissipating the doubt in
which Albert of Saxony had remained enveloped, da Vinci had declared
the velocity acquired by a falling body to be proportional to the time occu-

pied by the fall, but he did not know how to determine that law connecting the time consumed in falling with the space passed over by the falling body. Nevertheless, to find this law, it would have sufficed to invoke the following proposition: In a uniformly varied motion, the space traversed by the moving body is equal to that which it would traverse in a uniform motion, whose duration would be that of the preceding motion and whose velocity would be the same as that which affected the preceding motion at the mean instant of its duration. This proposition was known to Oresme, who had demonstrated it exactly as it was to be demonstrated later by Galileo. It was enunciated and discussed at the close of the fourteenth century by all the logicians who, in the University of Oxford, composed the school of William of Heytesbury, Chancellor of Oxford in 1375. It was subsequently examined or invoked in the fifteenth century by all the Italians who became the commentators of these logicians. And finally, the masters of the University of Paris, contemporaries of da Vinci, taught and demonstrated it as Oresme had done.

This law, which da Vinci was not able to determine, was published in 1545 by a Spanish Dominican, Domingo de Soto (1494–1560), an alumnus of the University of Paris and professor of theology at Alcala de Henares and afterward at Salamanca. He formulated the two laws thus:

The velocity of a falling body increases proportionally to the time of the fall.

The space traversed in a uniformly varied motion is the same as in a uniform motion occupying the same time, its velocity being the mean velocity of the former.

In addition, de Soto declared that the motion of a body thrown vertically upward is uniformly retarded. It should be mentioned that all these propositions were formulated by the celebrated Dominican as if in relation to truths generally admitted by the masters among whom he lived.

The Parisian theory, maintaining that the accelerated fall of bodies was caused by the effect of a continual increase of impetus caused by gravity, was admitted by Julius Caesar Scaliger (1484–1558), Benedetti, and Gabriel Vasquez (1551–1604), the celebrated Jesuit theologian. The first of these authors presented this theory in such a way that uniform acceleration of motion seemed naturally to follow from it.

De Soto, Tartaglia, and Cardano made strenuous efforts, after the manner of da Vinci, to explain the motion of projectiles by appealing to the conflict between impetus and gravity, but their attempts were frustrated by a peripatetic error which several Parisian masters had long before rejected. They believed that the motion of the projectile was accelerated from the start and attributed this initial acceleration to an impulse communicated by the vibrating air. Indeed, throughout the sixteenth century, the Italian Averroists continued to attribute to the ambient air the very transportation of the projectile. Tartaglia empirically discovered that a piece of artillery attained its greatest range when pointed at an angle of forty-five degrees to the horizon. Bruno insisted on Oresme's explanation of the fact that a body appears to fall in a vertical line in spite of the Earth's motion; to obtain the trajectory of this body, it is necessary to combine the action of its weight with the impetus which the Earth has imparted to it. It was in this way that Benedetti set forth the law followed by such an impetus. A body whirled in a circle and suddenly left to itself will move in a straight line tangent to the circle at the very point where the body happened to be at the moment of its release. For this achievement, Benedetti deserves to be ranked among the most valuable contributors to the discovery of the law of inertia. In 1553, Benedetti advanced the following argument: In air, or any fluid whatever, ten equal stones fall with the same velocity as one of their number, and if all were combined, they would still fall with the same velocity; therefore, in a fluid, two stones, one of which is ten times heavier than the other, fall with the same velocity. Benedetti lauded the extreme novelty of this argument with which, in reality, many scholastics had been familiar but which they had all claimed was not conclusive because the resistance which the air offered to the heavier stone could certainly not be ten times that which it opposed to the lighter one. Achillini was one of those who clearly maintained this principle. So that it might lead to a correct conclusion, Benedetti's argument had to be restricted to the motion of bodies in a vacuum, and this is what was done by Galileo.

XVII. Galileo's Work

Galileo Galilei (1564–1642) had been in youth a staunch peripatetic, but he was later converted to the Copernican system and devoted most of his efforts to its defense. The triumph of the system of Copernicus could be secured only by the perfecting of mechanics, and especially by solving the problem presented by the fall of bodies when the Earth was supposed to be in motion. It was toward this solution that many of Galileo's researches were directed, and to bring his labors to a successful issue, he had to adopt certain principles of Parisian dynamics. Unfortunately, instead of using them all, he left it to others to exhaust their fecundity.

Galilean statics was a compromise between the incorrect method inaugurated in Aristotle's *Mechanical Questions* and the correct method of virtual displacements successfully applied by the School of Jordanus. Imbued with ideas that were still intensely peripatetic, it introduced the consideration of a certain *impeto* or *momento*, proportional to the velocity of the moving body and not unlike the impetus of the Parisians. Galilean hydrostatics also showed an imperfect form of the principle of virtual displacements, which seemed to have been suggested to the great Pisan by the effectual researches made on the theory of running water by his friend Benedetto Castelli, the Benedictine (1577–1644). At first Galileo asserted that the velocity of a falling body increased proportionally to the space traversed; afterward, by an ingenious demonstration, he proved the utter absurdity of such a law. He then taught that the motion of a freely falling body was uniformly accelerated; in favor of this law, he contented himself with appealing to its simplicity without considering the continual increase of impetus under the influence of gravity. Gravity creates, in equal periods, a new and uniform impetus which, added to that already acquired, causes the total impetus to increase in arithmetical progression according to the time occupied in the fall; hence, the velocity of the falling body. This argument, toward which all Parisian tradition had been tending and which, in the last place, had been broached by Scaliger, leads to our modern law: A constant force produces uniformly accelerated motion. In Galileo's work, there is no trace either of the argument or of the conclusion deduced therefrom; however, the argument itself was carefully developed by Galileo's friend, Giovanni Battista Baliani (1582–1666).

From the very definition of velocity, Baliani endeavored to deduce the law according to which the space traversed by a falling body is increased proportionally to the time occupied in the fall. Here he was confronted by a difficulty that had also baffled da Vinci; however, he eventually anticipated its solution, which was given, after similar hesitation, by another of Galileo's disciples, Pierre Gassendi (1592–1655). Galileo had reached the law connecting the time occupied in the fall with the space traversed by a falling body by using a demonstration that became celebrated as the "demonstration of the triangle." It was literally what was given by Oresme in the fourteenth century; and, as we have seen, De Soto had thought of using Oresme's proposition in the study of the accelerated fall of bodies. Galileo extended the laws of freely falling bodies to a fall down an inclined plane and subjected to the test of experiment the law of the motion of a weight on an inclined plane.

A body which, without friction or resistance of any kind, would describe the circumference of a circle concentric with the Earth would retain an invariable *impeto* or *momento*, as gravity would in no way tend to increase or destroy this *impeto*. This principle, which belonged to the dynamics of Buridan and Albert of Saxony, was acknowledged by Galileo. On a small surface, a sphere concentric with the Earth is apparently merged into a horizontal plane; a body thrown on a horizontal plane and free from all friction would therefore assume a motion apparently rectilinear and uniform. It is only under this restricted and erroneous form that Galileo recognized the law of inertia, and in this, he was the faithful disciple of the School of Paris.

If a heavy body that is moved by an *impeto* which would make it describe a circle concentric with the Earth is, moreover, free to fall, the *impeto* of uniform rotation and gravity are component forces. To a small extent, the motion produced by this *impeto* may be assumed to be rectilinear, horizontal, and uniform; hence, the approximate law may be enunciated as follows: A heavy body, to which a horizontal initial velocity has been imparted at the very moment that it is abandoned to the action of gravity, assumes a motion which is sensibly the combination of a uniform horizontal motion with the vertical motion that it would assume without initial velocity. Galileo then demonstrated that the trajectory of this heavy body is a parabola with a vertical axis. This theory of the motion of projectiles rests on principles in no way conformable to an exact knowledge of the law of inertia and which are, at bottom, identical with those invoked by Oresme when he wished to explain how, despite the Earth's rotation, a body seems to fall vertically. The argument employed by Galileo did not permit him to state how a projectile moves when its initial velocity is not horizontal.

Evangelista Torricelli (1608–1647), a disciple of Castelli and Galileo, extended the latter's method to the case of a projectile whose initial velocity had a direction other than horizontal, and he proved that the trajectory remained a parabola with a vertical axis. On the other hand, Gassendi showed that in this problem of the motion of projectiles, the real law of inertia which had just been formulated by Descartes should be substituted for the principles admitted by the Parisian dynamics of the fourteenth century.

Mention should be made of Galileo's observations on the duration of the oscillation of the pendulum, as these observations opened up to dynamics a new field. Galileo's progress in dynamics served as a defense of the Copernican system, and the discoveries which, with the aid of the telescope, he was able to make in the heavens contributed to the same end. The spots on the sun's surface and the mountains (similar to those on the Earth), that hid from view certain portions of the lunar disc gave ample proof of the fact that the celestial bodies were not, as Aristotelian physics had maintained, formed of an incorruptible substance unlike sublunary elements. Moreover, the role of satellite which, in this heliocentric astronomy, the moon played in regard to the Earth was carried out in relation to Jupiter by the "Medicean planets," which Galileo had been the first to discover. Not satisfied with having defeated the arguments opposed to the Copernican system by adducing these excellent reasons, Galileo was eager to establish a positive proof in favor of this system. Inspired perhaps by Calcagnini, he believed that the phenomenon of the tides would furnish him the desired proof, and consequently he rejected every explanation of ebb and flow founded on the attraction of the sun and moon in order to attribute the motion of the seas to the centrifugal force produced by terrestrial rotation. Such an explanation would connect the period of high tide with the sidereal instead of the lunar day, thus contradicting the most ordinary and ancient observations. This remark alone ought to have held Galileo back and prevented him from producing an argument better calculated to overthrow the doctrine of the Earth's rotation than to establish and confirm it.

On two occasions, in 1616 and 1633, the Inquisition condemned what Galileo had written in favor of the system of Copernicus. The hypothesis of the Earth's motion was declared *falsa in Philosophia et ad minus erronea in fide*; the hypothesis of the sun being stationary was adjudged *falsa in Philosophia et formaliter haeretica*. Adopting the doctrine formulated by Tycho Brahe in 1578, the Holy Office forbade the use of all astronomical hypotheses that did not agree both with the principles of Aristotelian physics and with the letter of the Sacred Scriptures.

XVIII. Initial Attempts in Celestial Mechanics: Gilbert; Kepler

Copernicus had endeavored to describe accurately the motion of each of the celestial bodies, and Galileo had striven to show that the views of Copernicus were correct; but neither Copernicus nor Galileo had attempted to extend to the stars what they knew concerning the dynamics of sublunary motions, or to determine thereby the forces that sustain celestial motions. They were satisfied with holding that the daily rotation of the Earth is perpetuated by virtue of an impetus given once for all; that the various parts of an element belonging to a star tend toward the center of this star by reason of a gravity peculiar to each of the celestial bodies through which the body is enabled to preserve its entireness. Thus, in celestial mechanics, these two great scientists contributed scarcely any-thing to what had already been taught by Buridan, Oresme, and Nicholas of Cusa. About Galileo's time, we notice the first attempts to constitute celestial mechanics—that is, to explain the motion of the stars by the aid of forces analogous to those the effects of which we feel upon Earth. The most important of these initial attempts were made by William Gilbert (1540–1603) and Johann Kepler (1571–1631).

To Gilbert we are indebted for an exhaustive treatise on magnetism, in which he systematically incorporated what was known in medieval times of electrical and magnetic phenomena, without adding thereto anything very essential; he also gave the results of his own valuable experiments. It was in this treatise that he began to expound his *Magnetic Philosophy*—that is, his celestial mechanics—but the work in which he fully developed it was not published until 1651, long after his death. Like Oresme and Copernicus, Gilbert maintained that in each star there was a particular gravity through which the material parts belonging to this star—and these only—tended to rejoin the star when they had been separated from it. He compared this gravity, peculiar to each star, with the action by which a piece of iron flies toward the magnet whose nature it shares. This opinion, held by so many of Gilbert's predecessors and adopted by a great number of his imitators, led Francis Bacon astray. Bacon was the enthusiastic her-ald of the experimental method, which, however, he never practiced and of which he had an utterly false conception. According to Gilbert, the Earth, sun, and stars were animated, and the animating principle of each communicated to the body the motion of perpetual rotation. From a dis-tance, the sun exerted an action perpendicular to the radius vector which goes from the center of the sun to a planet, and this action caused the planet to revolve around the sun just as a horse turns the horse-mill to which it is yoked.

Kepler himself admitted that in his first attempts along the line of celestial mechanics, he was influenced by Nicholas of Cusa and by Gilbert. Inspired by the former of these authors, he attributed the Earth's rotation on its axis to an impetus communicated by the Creator at the beginning of time; but, under the influence of Gilbert's theory, he declared that this impetus ended by being transformed into a soul or an animating principle. In Kepler's earliest system, as in Gilbert's, the distant sun was said to exer-cise over each planet a power perpendicular to the radius vector, which power produced the circular motion of the planet. Kepler had the happy thought, however, of proposing a universal attraction for the magnetic attraction that Gilbert had considered peculiar to each star. He assumed that every material mass tended toward every other material mass, no matter to what celestial body each one belonged; that a portion of matter placed between two stars would tend toward the larger and nearer one, though it might never have belonged to it; that, at the moment of high tide, the waters of the sea rose toward the moon, not because they had any special affinity for this humid star, but by virtue of the general tendency that draws all material masses toward one another.

In the course of numerous attempts to explain the motion of the stars, Kepler was led to complicate his first celestial mechanics. He assumed that all celestial bodies were plunged into an aethereal fluid, that the rotation of the sun engendered a vortex within this fluid, the reactions of which interposed to deflect each planet from the circular path. He also thought that a certain power, similar to that which directs the magnetic needle, preserved invariable in space the direction of the axis around which the rotation of each planet is effected. The unstable and complicated system of celestial mechanics taught by Kepler sprang from deficient dynamics which, on many points, was more akin to that of the peripatetics than to that of the Parisians. These many vague hypotheses, however, exerted an incontestable influence on the attempts of scientists, from Kepler to Newton, to determine the forces that move the stars. If, indeed, Kepler prepared the way for Newton's work, it was mainly by the discovery of the three admirable laws that have immortalized his name; and by teaching that the planets described ellipses instead of circles, he produced in astronomy a revolution greater by far than that caused by Copernicus. He destroyed the last time-honored principle of ancient physics, according to which all celestial motions were reducible to circular motion.

XIX. Controversies Concerning Geostatics

The "magnetic" philosophy adopted and developed by Gilbert was not only rejected by Kepler but also badly abused in a dispute over the principles of statics. A number of the Parisian scholastics of the fourteenth century, and Albert of Saxony in particular, had accepted the principle that in every body there is a fixed, determined point which tends to join the center of the World, this point being identical to the center of gravity as considered by Archimedes. From this principle, various authors, notably da Vinci, deduced corollaries that retained a place in statics. The Copernican revolution had modified this principle but little, having simply substituted for the center of the universe a particular point in each star toward which tended the center of gravity of each mass belonging to this star. Copernicus, Galileo, and Gilbert admitted the principle thus modified, but Kepler rejected it. In 1635, Jean Beaugrand deduced from this principle a paradoxical theory on the gravity of bodies, and particularly on the variation in the weight of a body whose distance from the center of the universe changes. Opinions similar to those proposed by Beaugrand in his geostatics were held in Italy by Castelli and in France by Pierre Fermat (1608–1665). Fermat's doctrine was discussed and refuted by Etienne Pascal (1588–1651) and Gilles Persone de Roberval (1602–1675), and the admirable controversy between these authors and Fermat contributed in great measure to the clear exposition of a certain number of ideas employed in statics, among them that of the center of gravity.

It was this controversy which led Descartes to revive the question of virtual displacements in precisely the same form as that adopted by the School of Jordanus, in order that the essential propositions of statics might be given a stable foundation. On the other hand, Torricelli based all his arguments concerning the laws of equilibrium on the axiom quoted above: A system endowed with weight is in equilibrium when the center of gravity of all the bodies forming it is the lowest possible. Cardano and perhaps da Vinci had derived this proposition from the doctrine of Albert of Saxony, but Torricelli was careful to use it only under circumstances in which all verticals are considered parallel to one another, and, in this way, he severed all connection between the axiom that he admitted and the doubtful hypotheses of Parisian physics or magnetic philosophy. Thenceforth the principles of statics were formulated with accuracy; John Wallis (1616–1703), Pierre Varignon (1654–1722), and Jean Bernoulli (1667–1748) had merely to complete and develop the information provided by Stevin, Roberval, Descartes, and Torricelli.

XX. Descartes's Work

We have just stated what part Descartes took in the building of statics by bringing forward the method of virtual displacements, but his active interest in the building up of dynamics was still more important. He clearly formulated the law of inertia as observed by Benedetti: Every moving body is inclined, if nothing prevents it, to continue its motion in a straight line and with constant velocity; a body cannot move in a circle unless it is drawn toward the center by centripetal movement in opposition to the centrifugal force by which this body tends to fly away from the center. Because of the similarity of the views held by Descartes and Benedetti concerning this law, we may conclude that Descartes's discovery was influenced by that of

Benedetti, especially as Benedetti's works were known to Marin Mersenne (1588–1648), the faithful friend and correspondent of Descartes. Descartes connected the following truth with the law of inertia: A weight constant in size and direction causes a uniformly accelerated motion. Besides, we have seen how, with the aid of Descartes's principles, Gassendi was able to rectify what Galileo had taught concerning falling bodies and the motion of projectiles.

In statics, a heavy body can often be replaced by a material point placed at its center of gravity; but in dynamics, the question arises whether the motion of a body can be treated as if this body were entirely concentrated in one of these points, and also which point this is. This question relative to the existence and finding of a center of impulsion had already engrossed the attention of da Vinci and, after him, of Bernardino Baldi (1553–1617). Baldi asserted that in a body undergoing a motion of translation, the center of impulsion does not differ from the center of gravity. Now, is there a center of impulsion and, if so, where is it to be found in a body undergoing a motion other than that of translation—for instance, by rotation around an axis? In other words, is there a simple pendulum that moves in the same way as a given compound pendulum? Inspired, no doubt, by reading Baldi, Mersenne laid this problem before Roberval and Descartes, both of whom made great efforts to solve it but became unfriendly to each other because of the difference in their respective propositions. Of the two, Descartes came nearer to the truth, but the dynamic principles that he used were not sufficiently accurate to justify his opinion in a convincing manner; the glory was reserved for Christiaan Huygens.

The Jesuits, who at the College of La Flèche had been the preceptors of Mersenne and Descartes, did not teach peripatetic physics in its stereotyped integrity, but Parisian physics; the treatise that guided the instruction imparted at this institution being represented by the *Commentaries* on Aristotle, published by the Jesuits of Coimbra at the close of the seventeenth century. Hence, it can be understood why the dynamics of Descartes had many points in common with the dynamics of Buridan and the Parisians. Indeed, so close were the relations between Parisian and Cartesian physics that certain professors at La Flèche, such as Etienne Noël (1581–1660), became Cartesians. Other Jesuits attempted to build up a sort of combination of Galilean and Cartesian mechanics and the mechanics taught by Parisian scholasticism, and foremost among these men must be mentioned Honoré Fabri (1606–1688), a friend of Mersenne.

In every moving body, Descartes maintained the existence of a certain power to continue its motion in the same direction and with the same velocity; this power, which he called the *quantity of motion*, he measured by estimating the product of the mass of the moving body by the velocity that impels it. The affinity is close between the role which Descartes attributed to this quantity of motion and that which Buridan ascribed to impetus. Fabri was fully aware of this analogy, and the momentum that he discussed was at once the impetus of the Parisians and Descartes's quantity of motion. In statics, he identified this momentum with what Galileo called *momento* or *impeto*, and this identification was certainly conformable to the Pisan's idea. Fabri's synthesis was well adapted to make the truth clear that modern dynamics, the foundations of which were laid by Descartes and Galileo, proceeded almost directly from the dynamics taught during the fourteenth century in the University of Paris.

If the special physical truths demonstrated or anticipated by Descartes were easily traceable to the philosophy of the fourteenth century, the principles on which the great geometer wished to base these truths were absolutely incompatible with this philosophy. In fact, denying that in reality there existed anything qualitative, Descartes insisted that matter be reduced to extension and to the attributes of which extension seemed to him susceptible—namely, numerical proportions and motion. And it was by combinations of different figures and motions that all the effects of physics could be explained according to his liking. Therefore, the power by virtue of which a body tends to preserve the direction and velocity of its motion is not a quality distinct from motion, such as the impetus recognized by the scholastics; it is nothing other than the motion itself, as was taught by William of Ockham at the beginning of the fourteenth century. A body in motion and isolated would always retain the same quantity of motion, but there is no isolated body in a vacuum because, matter being identical to extension, vacuum is inconceivable, as is also compressibility. The only conceivable motions are those which can be produced in the midst of incompressible matter—that is, vortical motions confined within their own bulk.

In these motions, bodies drive one another from the place they have occupied, and in such a transmission of motion, the quantity of motion of each of these bodies varies; however, the entire quantity of motion of all bodies that impinge on one another remains constant, as God always maintains the same sum total of motion in the World. This transmission of motion by impact is the only action that bodies can exert over one another, and in Cartesian as well as Aristotelian physics, a body cannot put another in motion unless it touches it, immediate action at a distance being beyond conception.

There are various species of matter, differing from one another only in the size and shape of the contiguous particles of which they are formed. The space that extends between the different heavenly bodies is filled with a certain subtle matter, the fine particles of which easily penetrate the interstices left between the coarser constituents of other bodies. The properties of subtle matter play an important part in all Cartesian cosmology. The vortices in which subtle matter moves, and the pressure generated by these vortical motions, serve to explain all celestial phenomena. Leibniz was right in supposing that for this part of his work, Descartes had drawn largely on Kepler. Descartes also strove to explain, with the aid of the figures and motions of subtle and other matter, the different effects observable in physics, particularly the properties of the magnet and of light. Light is identical to the pressure which subtle matter exerts over bodies, and, as subtle matter is incompressible, light is instantly transmitted to any distance, however great.

The supposition by the aid of which Descartes attempted to reduce all physical phenomena to combinations of figures and motions had scarcely any part in the discoveries that he made in physics. Therefore, the identification of light with the pressure exerted by subtle matter plays no part in the invention of the new truths which Descartes taught in optics. Foremost among these truths is the law of the refraction of light passing from one medium to another, although the question still remains whether Descartes discovered this law himself or whether, as Huygens accused him of doing, he borrowed it from Willebrord Snellius (1591–1626), without any mention of the real author. By this law, Descartes gave the theory of refraction through a prism, which permitted him to measure the indices of refraction; moreover, he greatly perfected the study of lenses and finally completed the explanation of the rainbow, no progress having been made along this line from the year 1300, when Theodoric of Freiberg had given his treatise on it. The reason the rays emerging from the drops of water are variously colored, however, was no better known by Descartes than by Aristotle; it remained for Newton to make the discovery.

XXI. Progress of Experimental Physics

Even in Descartes's work, the discoveries in physics were almost independent of Cartesianism. The knowledge of natural truths continued to advance without the influence of this system and, at times, even in opposition to it, although those to whom this progress were due were often Cartesians. This advancement was largely the result of a more frequent and

skillful use of the experimental method. The art of making logically connected experiments and of deducing their consequences is indeed ancient;
in a way, the works produced by this art were no more perfect than the
researches of Pierre of Maricourt on the magnet or Theodoric of Freiberg
on the rainbow. However, if the art remained the same, its technic continued to improve; more skilled workmen and more powerful processes furnished physicists with more intricate and better-made instruments, and
thus rendered possible more delicate experiments. The imperfect tests
made by Galileo and Mersenne in endeavoring to determine the specific
weight of air mark the beginning of the development of the experimental
method, which was at once vigorously pushed forward by discussions
about vacuum.

In peripatetic physics, the possibility of an empty space was a logical
contradiction, but after the condemnation pronounced at Paris in 1277 by
Tempier, the existence of a vacuum ceased to be considered absurd. It was
simply taught as a fact that the powers of nature are so constructed as to
oppose the production of an empty space. Of the various conjectures proposed concerning the forces which prevent the appearance of a vacuum,
the most sensible and, it would seem, the most generally received among
sixteenth-century Parisians was the following: Contiguous bodies adhere
to one another, and this adhesion is maintained by forces resembling those
by which a piece of iron adheres to the magnet which it touches. In naming
this force *horror vacui*, there was no intention of considering the bodies as
animate beings. A heavy piece of iron detaches itself from the magnet that
should hold it up, its weight having conquered the force by which the
magnet retained it; in the same way, the weight of too heavy a body can
prevent the *horror vacui* from raising this body. This logical corollary of
the hypothesis we have just mentioned was formulated by Galileo, who
saw therein the explanation of a fact well known to the cistern makers of
his time; namely, that a suction pump could not raise water higher than
thirty-two feet. This corollary entailed the possibility of producing an
empty space, a fact known to Torricelli, who, in 1644, made the celebrated
experiment with mercury that was destined to immortalize his name. At
the same time, however, he anticipated a new explanation of this experiment: The mercury is supported in the tube not by the *horror vacui* that
does not exist but by the pressure which the heavy air exerts on the exterior surface of the basin.

Torricelli's experiment quickly attracted the attention of physicists. In
France, thanks to Mersenne, it called forth on his part, and on that of those
who had dealings with him, many experiments in which Roberval and Pas-

cal (1623–1662) vied with each other in ingenuity, and in order to have the resources of technic more easily at his disposal, Pascal made his startling experiments in a glass factory at Rouen. Among the numerous inquirers interested in Torricelli's experiment, some accepted the explanation offered by the "column of air" and advanced by the great Italian geometer himself; whereas others, such as Roberval, held to the ancient hypothesis of an attraction analogous to magnetic action. At length, with a view toward settling the difference, an experiment was made which consisted in measuring at what height the mercury remained suspended in Torricelli's tube, observing it first at the foot of a mountain and then at the summit. The idea of this experiment seemed to have suggested itself to several physicists, notably Mersenne, Descartes, and Pascal, and through the instrumentality of the last named and the courtesy of Périer, his brother-in-law, it was made between the base and summit of Puy-de-Dome, September 19, 1648. The *Traité de l'equilibre de liqueurs et de la pesanteur de la masse de l'air*, which Pascal subsequently composed, is justly cited as a model of the art of logically connected experiments with deductions. There were many discussions between atomists and Cartesians as to whether the upper part of Torricelli's tube was really empty or was filled with subtle matter, but these discussions bore little fruit. Fortunately for physics, however, the experimental method so accurately followed by Torricelli, Pascal, and their rivals continued to progress.

Otto von Guericke (1602–1686) seems to have preceded Torricelli in the production of an empty space, since, between 1632 and 1638, he appears to have constructed his first pneumatic machine; with the aid of this instrument, in 1654 he made the celebrated Magdeburg experiments, published in 1657 by his friend Gaspar Schott, S. J. (1608–1660). Informed by Schott of Guericke's researches, Robert Boyle (1620–1691) perfected the pneumatic machine and, assisted by Richard Townley, his pupil, pursued the experiments that made known the law of the compressibility of perfect gases. In France these experiments were taken up and followed by Mariotte (1620–1684). The use of the dilatation of a fluid for showing the changes of temperature was already known to Galileo, but it is uncertain whether the thermoscope was invented by Galileo or by one of the numerous physicists to whom the priority is attributed, among these being Santorio, called Sanctorius (1560–1636), Fra Paolo Sarpi (1552–1623), Cornelis van Drebbel (1572–1634), and Robert Fludd (1574–1637). Although the various thermoscopes for air or liquid used in the beginning admitted of only arbitrary graduation, they nevertheless served to indicate the constancy of the temperature or the direction of its variations and consequently contributed to the discovery of a number of laws of physics. Hence, this apparatus was used in the Accademia del Cimento, which was opened at Florence on June 19, 1657, and was devoted to the study of experimental physics. To the members of this academy we are especially indebted for the demonstration of the constancy of the point of fusion of ice and of the absorption of heat accompanying this fusion. Observations of this kind, made by means of the thermoscope, created an ardent desire for the transformation of this apparatus into a thermometer, by the aid of a definite graduation so arranged that instruments could be made everywhere which would be comparable with one another. This problem, one of the most important in physics, was not solved until 1702, when Guillaume Amontons (1663–1705) worked it out in the most remarkable manner. Amontons took as a starting point these two laws, discovered or verified by him: The boiling point of water under atmospheric pressure is constant. The pressures sustained by any two masses of air, heated in the same way in any two constant volumes, have a relation independent of the temperature. These two laws enabled Amontons to use the air thermometer under constant volume and to graduate it in such a way that it gave what we today call absolute temperature. Of all the definitions of the degree of temperature given since Amontons's time, he, at the first stroke, found the most perfect. Equipped with instruments capable of measuring pressure and registering temperature, experimental physics could not but make rapid progress, this being still further augmented by reason of the interest shown by the learned societies that had recently been founded. The Accademia del Cimento was discontinued in 1667, but the Royal Society of London had begun its sessions in 1663, and the Académie des Sciences at Paris was founded or, rather, organized by Colbert in 1666. These academies immediately became the enthusiastic centers of scientific research in regard to natural phenomena.

XXII. Undulatory Theory of Light

It was to the Académie des Sciences of Paris that, in 1678, Christiaan Huygens (1629–1695) presented his *Treatise on Light*. According to the Cartesian system, light was instantly transmitted any distance through the medium of incompressible subtle matter. Descartes did not hesitate to assure Fermat that his entire philosophy would give way as soon as it could be demonstrated that light is propagated with a limited velocity. In 1675, Ole Römer (1644–1710), the Danish astronomer, announced to the

Académie des Sciences the extent of the considerable but finite velocity with which light traverses the space that separates the planets from one another; the study of the eclipses of Jupiter's satellites had brought him to this conclusion. Descartes's optical theory was destroyed, and Huygens undertook to build a new theory of light. He was constantly guided by the supposition that, in the midst of compressible aether, substituted for incompressible subtle matter, light is propagated by waves exactly similar to those which transmit sound through a gaseous medium. This comparison led him to an explanation, which is still the standard one, of the laws of reflection and refraction. In this explanation, the index of the refraction of light passing from one medium to another equals the ratio of the velocity of propagation in the first medium to the velocity of propagation in the second. In 1850, this fundamental law was confirmed by Foucault's experiments.

Huygens did not stop here, however. In 1669, Erasmus Bartolin, known as Bartholinus (1625–1698), discovered the double refraction of Iceland spar. By using a generalization, as ingenious as it was daring, of the theory he had given for noncrystallized media, Huygens succeeded in tracing the form of the surface of a luminous wave inside a crystal such as spar or quartz, and in defining the apparently complex laws of the double refraction of light in the interior of these crystals. At the same time, he called attention to the phenomena of polarization which accompany this double refraction; he was, however, unable to draw from his optical theory the explanation of these effects. The comparison between light and sound caused Malebranche (1638–1715) to make some effective conjectures in 1699. He assumed that light is a vibratory motion analogous to that produced by sound; the greater or lesser amplitude of this motion, as the case may be, generates a greater or lesser intensity, but while in sound each period corresponds to a particular note, in light it corresponds to a particular color. Through this analogy, Malebranche arrived at the idea of monochromatic light, which Newton was to deduce from admirably conducted experiments; moreover, he established between simple color and the period of the vibration of light the connection that was to be preserved in the optics of Young and Fresnel.

XXIII. Developments of Dynamics

Both Cartesians and atomists maintained that impact was the only process by which bodies could put one another in motion: hence, to Cartesians and atomists, the theory of impact seemed like the first chapter of rational physics. This theory had already enlisted the attention of Galileo, Marcus Marci (1639), and Descartes when, in 1668, the Royal Society of London proposed it as the subject of a competition; of the three important memoirs submitted to the criticism of this society by John Wallis, Christopher Wren (1632–1723), and Huygens, the last is the only one that we can consider. In his treatise, Huygens adopted the following principle: If a material body, subject merely to the action of gravity, starts from a certain position with initial velocity equal to zero, the center of gravity of this body can at no time rise higher than it was at the outset of the motion. Huygens justified this principle by observing that if it were false, perpetual motion would be possible. To find the origin of this axiom, it would be necessary to go back to *De subtilitate* by Cardano, who had probably drawn it from the notes of da Vinci; the proposition on which Torricelli had based his statics was a corollary from this postulate. By maintaining the accuracy of this postulate, even in the case in which parts of the system clash; by combining it with the law of the accelerated fall of bodies, taken from Galileo's works, and with another postulate on the relativity of motion, Huygens arrived at the law of the impact of hard bodies. He showed that the quantity the value of which remains constant in spite of this impact is not, as Descartes declared, the total quantity of motion but that which Leibniz called the quantity of *vis viva* (living force).

The axiom that had so happily served Huygens in the study of the impact of bodies he now extended to a body oscillating around a horizontal axis, and his *Horologium oscillatorium*, which appeared in 1673, solved in the most elegant and complete manner the problem of the centers of oscillation previously handled by Descartes and Roberval. That Huygens's axiom was the subversion of Cartesian dynamics was shown by Leibniz in 1686. If, like Descartes, we measure the efficiency of a force by the work that it does, and if, moreover, we admit Huygens's axiom and the law of falling bodies, we find that this efficiency is not measured by the increase in the quantity of motion of the moving body but by the increase in half the product of the mass of the moving body and the square of its velocity. It was this product that Leibniz called *vis viva*. Huygens's *Horologium oscillatorium* gave not only the solution of the problem of the center of oscillation but likewise a statement of the laws which, in circular motion, govern the magnitude of centrifugal force, and thus it was that the eminent physicist prepared the way for Newton, the lawgiver of dynamics.

XXIV. Newton's Work

Most of the great dynamical truths had been discovered between the time of Galileo and Descartes and that of Huygens and Leibniz. The science of dynamics required a Euclid who would organize it as geometry had been organized, and this Euclid appeared in the person of Isaac Newton (1642–1727) who, in his *Philosophiae naturalis principia mathematica*, published in 1687, succeeded in deducing the entire science of motion from three postulates: inertia, the independence of the effects of previously acquired forces and motions, and the equality of action and reaction. Had Newton's *Principia* contained nothing more than this coordination of dynamics into a logical system, they would nevertheless have been one of the most important works ever written; but, in addition, they gave the grandest possible application of this dynamics in utilizing it for the establishment of celestial mechanics. In fact, Newton succeeded in showing that the laws of bodies falling to the surface of the Earth, the laws that preside over the motion of planets around the sun and of satellites around the planets which they accompany, and, finally, the laws that govern the form of the Earth and of the other stars, as also the high and low tides of the sea, are but so many corollaries from this unique hypothesis: Two bodies, whatever their origin or nature, exert over each other an attraction proportional to the product of their masses and in inverse ratio to the square of the distance that separates them.

The dominating principle of ancient physics declared the essential distinction between the laws that directed the motions of the stars—beings exempt from generation, change, and death—and the laws presiding over the motions of sublunary bodies subject to generation and corruption. From the birth of Christian physics and especially from the end of the thirteenth century, physicists had been endeavoring to destroy the authority of this principle and to render the celestial and sublunary worlds subject to the same laws, the doctrine of universal gravitation being the outcome of this prolonged effort. In proportion as the time approached when Newton was to produce his system, attempts at cosmology were multiplied, so many forerunners, as it were, of this discovery. When, in 1672, Guericke again took up Kepler's celestial mechanics, he made only one correction therein, which unfortunately caused the disappearance of the only proposition by which this work led up to Newton's discoveries. Kepler had maintained that two material masses of any kind attract each other, but in imitation of Copernicus, Gilbert, and Galileo, Guericke limited this mutual attraction to parts of the same star so that, far from being attracted by the Earth, portions of the moon would be repelled by the Earth if placed on its surface. But in 1644, under the pseudonym Aristarchus of Samos, Roberval published a system of celestial mechanics in which the attraction was perhaps mutual between two masses of no matter what kind; in which, at all events, the Earth and Jupiter attracted their satellites with a power identical to the gravity with which they endowed their own fragments. In 1665, on the pretense of explaining the motions of Jupiter's satellites, Giovanni Alfonso Borelli (1608–1679) tried to advance a theory which simultaneously comprised the motions of the planets around the sun and of the satellites around the planets. He was the first of modern scientists (Plutarch having preceded him) to hold the opinion that the attraction which causes a planet to tend toward the sun and a satellite to tend toward the star which it accompanies is in equilibrium with the centrifugal force produced by the circular motion of the planet or satellite in question. In 1674, Robert Hooke (1635–1702) formulated the same idea with great precision. Having already supposed the attraction of two masses to vary inversely as the square of their distance, he possessed the fundamental hypotheses of the theory of universal gravitation, which hypotheses were held by Wren about the same time. Neither of these scientists, however, was able to deduce therefrom celestial mechanics, as both were still unacquainted with the laws of centrifugal force, published at this time by Huygens. In 1684, Edmund Halley (1656–1742) strove to combine Huygens's theories with Hooke's hypotheses, but before his work was finished, Newton presented his *Principia* to the Royal Society, having for twenty years silently pursued his meditations on the system of the World. Halley, who could not forestall Newton, had the glory of broadening the domain of universal gravitation by making it include comets (1705).

Not satisfied with creating celestial mechanics, Newton also contributed largely to the progress of optics. From ancient times, the coloring of the spectrum, produced by the passage of white light through a glass prism, had elicited the wonder of observers and appealed to the acumen of physicists without, however, being satisfactorily explained. Finally, a complete explanation was given by Newton, who, in creating a theory of colors, accomplished what all the philosophers from Aristotle down had labored in vain to achieve. The theory advanced by the English physicist agreed with that proposed by Malebranche at the same time. Malebranche's theory, however, was nothing more than a hypothesis suggested by the analogy between light and sound, whereas Newton's explanation was drawn from experiments, as simple as they were ingenious, its exposition by the author was one of the most beautiful examples of experimental induction. Unfortunately, Newton disregarded this analogy between sound and light that had furnished Huygens and Malebranche with such fruitful discoveries. Newton's opinion was to the effect that light is formed of infinitely small projectiles thrown off with extreme velocity by incandescent bodies. The particles of the medium in which these projectiles move exert over them an attraction similar to universal attraction; however, this new attraction does not vary inversely as the square of the distance but according to another function of the distance, and in such a way that it exercises a great power between a material particle and a luminous corpuscle that are contiguous. Nevertheless, this attraction becomes altogether insensible as soon as the two masses between which it operates are separated from each other by a perceptible interval.

This action exerted by the particles of a medium on the luminous corpuscles pervading them changes the velocity with which these bodies move and the direction they follow at the moment of passing from one medium to another; hence the phenomenon of refraction. The index of refraction is the ratio of the velocity of light in the medium which it enters to the velocity it had in the medium which it leaves. Now, as the index of refraction so understood was precisely the reverse of that attributed to it by Huygens's theory, in 1850, Foucault submitted both to the test of experiment, with the result that Newton's theory of emission was condemned. Newton explained the experimental laws that govern the coloring of thin laminae, such as soap bubbles, and succeeded in compelling these colors, by suitable forms of these thin laminae, to assume the regular order known as "Newton's Rings." To explain this phenomenon, he conceived that luminous projectiles have a form that may, at the surface of contact of two media, either pass easily or be easily reflected, according to the manner of their presentation at the moment of passage; a rotary motion causes them to pass alternately by "fits of easy transmission or of easy reflection."

Newton thought that he had accounted for the principal optical phenomena by supposing that, besides this universal attraction, there existed an attraction, sensible only at a very short distance, exerted by the particles of bodies on luminous corpuscles, and naturally he came to believe that these two kinds of attraction would suffice to explain all physical phenomena. Action extending to a considerable distance, such as electric and magnetic action, must follow laws analogous to those which govern universal gravity; on the other hand, the effects of capillarity and cohesion, chemical decomposition, and reaction must depend on molecular attraction extending only to extremely small distances and similar to that exerted over luminous corpuscles. This comprehensive hypothesis proposed by Newton in a "question" placed at the end of the second edition of his *Optics* (1717) gave a sort of outline of the program which eighteenth-century physics was to attempt to carry out.

XXV. Progress of General and Celestial Mechanics in the Eighteenth Century

This program made three demands: first, that general mechanics and celestial mechanics advance in the manner indicated by Newton; second, that electric and magnetic phenomena be explained by a theory analogous to that of universal gravitation; and third, that molecular attraction furnish the detailed explanations of the various changes investigated by physics and chemistry.

Many followed in the path outlined by Newton and tried to extend the domain of general and celestial mechanics, but there were three who seem to have surpassed all the others: Alexis-Claude Clairaut (1713–1765), Jean-Baptiste le Rond d'Alembert (1717–1783), and Leonhard Euler (1707–1783). The progress which, thanks to these three able men, was made in general mechanics may be summed up as follows. In 1743, by his principle of the equilibrium of channels, which was easily connected with the principle of virtual displacements, Clairaut obtained the general equations of the equilibrium of liquids. In the same year, d'Alembert formulated a rule whereby all problems of motion were reduced to problems of equilibrium and, in 1744, applied this rule to the equation of hydrostatics given by Clairaut and arrived at the equations of hydrodynamics. Euler transformed these equations and, in his studies on the motion of liquids, was able to obtain results no less important than those which he had obtained by analyzing the motion of solids. Clairaut extended the consequences of universal attraction in all directions, and in 1743, the equations of hydrostatics that he had established enabled him to perfect the theory of the figure of the earth. In 1752, he published his theory of lunar inequalities, which he had at first despaired of accounting for by Newton's principles. The methods that he devised for the study of the perturbations which the planets produce on the path of a star permitted him, in 1758, to announce with accuracy the time of the return of Halley's Comet. The confirmation of this prediction in which Clairaut had received assistance from Lalande (1732–1807) and Mme. Lepaute, both able mathematicians, placed beyond doubt the applicability of Newton's hypotheses to comets.

Great as were Clairaut's achievements in perfecting the system of universal attraction, they were not as important as those of d'Alembert. Newton could not deduce from his suppositions a satisfactory theory of the precession of the equinoxes, and this failure marred the harmony of the doctrine of universal gravitation. In 1749, d'Alembert deduced from the hypothesis of gravitation the explanation of the precession of the equinoxes and of the nutation of the earth's axis; and soon afterward, Euler, drawing on the admirable resources of his mathematical genius, made still further improvements on d'Alembert's discovery. Clairaut, d'Alembert, and Euler were the most brilliant stars in an entire constellation of mechanical theorists and astronomers, and after this group came another, in which shone two men of surpassing intellectuality: Joseph-Louis Lagrange (1736–1813) and Pierre-Simon Laplace (1749–1827). Laplace was said to have been born to complete celestial mechanics, if, indeed, it were in the nature of a science to admit of completion; and quite as much could be said of Lagrange with regard to general mechanics. In 1787, Lagrange published the first edition of his *Mécanique analytique*; the second, which was greatly enlarged, was published after the author's death. Laplace's *Mécanique celeste* was published from 1799 to 1805, and both of these works give an account of the greater part of the mechanical conquests made in the course of the eighteenth century, with the assistance of the principles that Newton had assigned to general mechanics and the laws that he had imposed on universal gravitation. However exhaustive and effective these two treatises are, they do not by any means include all the discoveries in general and celestial mechanics for which we are indebted to their authors. To do Lagrange even meager justice, his able researches should be placed on a par with his *Mécanique analytique*; and our idea of Laplace's work would be very incomplete were we to omit the grand cosmogonic hypothesis with which, in 1796, he crowned his *Exposition du système du monde*. In developing this hypothesis, the illustrious geometer was unaware that in 1755, Kant had expressed similar suppositions which were marred by serious errors in dynamic theories.

XXVI. Establishment of the Theory of Electricity and Magnetism

For a long time, the study of electric action was merely superficial, and at the beginning of the eighteenth century, it was still in the condition in which Thales of Miletus had left it, remaining far from the point to which the study of magnetic attraction and repulsion had been carried in the time of Pierre de Maricourt (Petrus Peregrinus). When, in 1733 and 1734, Charles-François de Cisternay du Fay distinguished two kinds of electricity, resinous and vitreous, and when he proved that bodies charged with the same kind of electricity repel one another, whereas those charged with different kinds attract one another, electrical science was brought to the level that magnetic science had long before attained, and thenceforth these two sciences, united by the closest analogy, progressed side by side. They advanced rapidly as, in the eighteenth century, the study of electrical phenomena became a popular craze. Physicists were not the only ones devoted to it; men of the world crowded the salons where popularizers of the science, such as the Abbé Nollet (1700–1770), enlisted as votaries dandified marquesses and sprightly marchionesses. Numerous experimentalists applied themselves to multiplying observations on electricity and magnetism, but we shall restrict ourselves to mentioning Benjamin Franklin (1706–1790), who, by his logically conducted researches, contributed more than any other man to the formation of the theories of electricity and magnetism. The researches of Henry Cavendish (1731–1810) deserve to be placed in the same rank as Franklin's, though they were little known before his death.

By means of Franklin's experiments and his own, Aepinus (Franz Ulrich Theodor Hoch, 1724–1802) became the first to attempt to solve the problem suggested by Newton and, by the hypothesis of attractive and repellent forces, to explain the distribution of electricity and magnetism over the bodies which they affect. His researches could not be pushed very far, as it was still unknown that these forces depend on the distance at which they are exerted. Moreover, Aepinus succeeded in drawing still closer the connection already established between the sciences of electricity and magnetism, by showing the polarization of each of the elements of the insulating plate which separates the two collecting plates of the condenser. The experiment he made in this line in 1759 was destined to suggest to Coulomb the experiment of the broken magnets and the theory of magnetic polarization, which is the foundation of the study of magnets; it was also to be the starting point of an entire branch of electrical science— namely, the study of dielectric bodies, which study was developed in the nineteenth century by Michael Faraday and James Clerk Maxwell.

Their analogy to the fertile law of universal gravitation undoubtedly led physicists to suppose that electrical and magnetic forces vary inversely as the square of the distance that separates the acting elements; but so far, this opinion had not been confirmed by experiment. In 1780, however, it received this confirmation from Charles Augustin Coulomb with the aid of the torsion balance. By the use of this balance and the proof plane, he was able to make detailed experiments on the subject of the distribution of electricity over conductive bodies; no such tests had previously been made. Although Coulomb's experiments placed beyond doubt the elementary laws of electricity and magnetism, it still remained to be established by mathematical analysis how electricity was distributed over the surface of conductive bodies of given shape and how a piece of soft iron was magnetized under given circumstances. The solution of these problems was attempted by Coulomb and also, in 1787, by Hauy, but neither of these two savants pushed his tests very far. The establishment of principles which would permit analysis of the distribution of electricity on conductors and of magnetism on soft iron required the genius of Simon-Denis Poisson (1781–1840).

In 1812, Poisson showed how the investigation of the distribution of electricity in equilibrium on conductors belonged to the domain of analysis, and he gave a complete solution of this problem in the case of two conductive spheres influencing each other, whether placed at given distances or in contact. Coulomb's experiments in connection with contiguous spheres established the truth of Poisson's theory. In 1824, Poisson established, based on the subject of hollow conductors limited either internally or externally by a spherical cavity, theorems which, in 1828, were extended by George Green (1793–1841) to all kinds of hollow conductors and which Faraday was subsequently to confirm through experimentation. Between 1813 and 1824, Poisson took up the study of magnetic forces and magnetization by impulsion; in spite of a few inaccuracies which the future was to correct, the formulae which he established remain at the basis of all the research of which magnetism has meanwhile been the object. Thanks to Poisson's memoirs, the theory of the forces exercised in inverse ratio to the square of the distance, by annexing the domain of static electricity and magnetism, markedly enlarged the field which at first included only celestial mechanics. The study of the action of the electric current was to open up to this theory a new and fertile territory.

The discoveries of Luigi Galvani (1737–1798) and Alessandro Volta (1745–1827) enriched physics with the voltaic battery. It would be impossible to enumerate, even briefly, the researches occasioned by this discovery. All physicists have compared the conductor, the seat of a current, to a space in which a fluid circulates. In his works on hydrodynamics, Euler had established general formulae which apply to the motion of all fluids; and, imitating Euler's method, Jean Baptiste Joseph Fourier (1768–1830) began the study of the circulation of heat—then considered a fluid and called *caloric*—within conductive bodies. The mathematical laws to which he had recourse once more showed the extreme importance of the mathematical methods inaugurated by Lagrange and Laplace in the study of universal attraction and at the same time extended by Poisson to the study of electrostatics. In order to treat mathematically the circulation of electric fluid in the interior of conductive bodies, it sufficed to take up Fourier's analysis almost literally, substituting the word *electricity* for the word *heat*; this was done in 1827 by Georg Simon Ohm (1789–1854).

Meanwhile, on July 21, 1820, Hans Christian Oersted (1777–1851) discovered the action of the electric current on the magnetic needle. To this discovery André-Marie Ampère (1775–1836) added that of the action exerted over each other by two conductors carrying electric currents, and to the study of electrodynamic and electromagnetic forces, he applied a method similar to that used by Newton when studying universal attraction. In 1826, Ampère gave the complete theory of all these forces in his *Mémoire sur la théorie mathématique des phénomènes électro-dynamiques uniquement déduite de l'expérience*, a work that can stand the test of comparison with the *Philosophiae naturalis principia mathematica* and not be found wanting.

Not wishing to carry the history of electricity and magnetism beyond this date, we shall content ourselves with making another comparison between the two works we have just mentioned. As Newton's treatise brought about numerous discoveries on the part of his successors, Ampère's memoir gave the initial impetus to researches which have greatly broadened the field of electrodynamics and electromagnetism. Michael Faraday (1791–1867), an experimentalist whose activity, skill, and good fortune have perhaps never been equaled, established in 1831 the experimental laws of electrodynamic and electromagnetic induction; and between 1845 and 1847, Franz Ernst Neumann (1798–1895) and Wilhelm Weber (1804–1891), by closely following Ampère's method of studying electrodynamic force, finally established the mathematical theory of these phenomena of induction. Michael Faraday was opposed to Newtonian doctrines and highly disapproved of the theory of action at a distance; in fact, when he applied himself to analyzing the polarization of insulated media, which he called dielectrics, he hoped to eliminate the hypothesis of such action. Meantime, by extending to dielectric bodies the formulae that Poisson, Ampère, and Neumann had established for magnets and conductive bodies, James Clerk Maxwell (1831–1879) was able to create a new branch of electrodynamics and thereby bring to light the long-sought link connecting the sciences of electricity and optics. This wonderful discovery was not one of the least important conquests of the method defined and practiced by Newton.

XXVII. Molecular Attraction

While universal attraction, which varies proportionally as the product of the masses and inversely as the square of the distance, was being established throughout the science of astronomy, and while, thanks to the study of other forces also varying inversely as the square of the distance, electricity and magnetism were being organized, other parts of physics received no less light from another Newtonian hypothesis—namely, the supposition that between two material particles, there is an extremely powerful attraction distinct from universal attraction while the two particles are contiguous but which ceases to be appreciable as soon as the two masses which it acts on are separated by a perceptible distance. Among the phenomena to be explained by such attractions, Newton had already signalized the effect of capillarity, in connection with which Francis Hauksbee (d. 1705) had made interesting experiments. In 1718, James Jurin (1684–1750) tried to follow Newton's idea but without any marked success, and it was Clairaut who, in 1743, showed how hydrostatic methods permitted the application of this idea to the explanation of capillary phenomena. Unfortunately, his able reasoning led to no important result, as he had ascribed too great a value to the extent of molecular action.

Chemical action was also one of the actions which Newton made subject to molecular attraction, and John Keill (1671–1721), John Freind (1675–1728), and Pierre Joseph Macquer (1718–1784) believed in the fruitfulness of this Newtonian opinion. The hypothesis of molecular attraction proved a great annoyance to a man whose scientific mediocrity had not prevented him from acquiring great influence—we mean Georges-Louis-Leclerc, Comte de Buffon (1707–1788). Incapable of understanding that an attraction could be other than inversely proportional to the square of the distance, Buffon entered into a discussion of the subject with Clairaut and fondly imagined that he had triumphed over the modest learning of his opponent. Roger Joseph Boscovich, S.J. (1711–1787) published a detailed exposition of the views attacked by Buffon and defended by Clairaut, and, inspired alike by the opinions of Newton and Leibniz, he conceived a cosmology in which the universe is composed solely of material points, which are attracted to each other in pairs. When these points are separated by a perceptible distance, their attraction is reduced to mere universal attraction, whereas when they are in very close proximity, it assumes a dominant importance. Boscovich's cosmology provided physical theory with a program which the geometers of the eighteenth century, and of a great portion of the nineteenth, labored assiduously to carry out.

The efforts of Johann Andreas von Segner (1704–1777), and subsequently of Thomas Young (1773–1829), again drew attention to capillary phenomena, and with the assistance of the hypothesis of molecular attraction, as also of Clairaut's method, Laplace advanced in 1806 and 1807 an admirable theory, which Karl Friedrich Gauss (1777–1855) improved in 1829. Being a thoroughly convinced partisan of Boscovich's cosmological doctrine, Laplace communicated his convictions to numerous geometers, who surrendered to the ascendancy of his genius; we shall only mention Claude-Louis-Marie Navier (1785–1836), Poisson, and Augustin Cauchy (1789–1857). In developing the consequences of the hypothesis of molecular attraction, Navier, Poisson, and Cauchy succeeded in building the theory of the equilibrium and small motions of elastic bodies, one of the finest and most fruitful theories of modern physics. The discredit into which the progress of present-day thermodynamics has brought Boscovich's cosmology has, however, affected scarcely anything of what Laplace, Gauss, Navier, Poisson, Cauchy, and many others have deduced from the principles of this cosmology. The theories they established have always been readily justified with the assistance of new methods, the way of bringing about this justification having been indicated by Cauchy himself and George Green. After Macquer, many chemists used the hypothesis of molecular attraction in an attempt to disentangle the laws of reaction which they studied, and among these scientists we may mention Torbern Bergman (1735–1784) and, above all, Claude Louis Berthollet (1784–1822). When the latter published his *Statique chimique* in 1803, he believed that the science of chemical equilibria, subject at last to Newton's method, had found its true direction; however, it was not to take this direction until much later, when it would be guided by precepts altogether different and which were to be formulated by thermodynamics.

XXVIII. Revival of the Undulatory Theory of Light

The emission theory of light not only led Newton to conceive the hypothesis of molecular attraction but also seemed to provide this hypothesis with an opportunity for further success by permitting Laplace to find, in the emission system, the laws of the double refraction of Iceland spar, which laws Huygens had discovered by the use of the undulatory theory. In this way, Newton's optics appeared to rob Huygens's optics of the one advantage in which it gloried. At the very moment that Laplace's discovery seemed to ensure the triumph of the emission system, however, the undulatory theory carried off new and dazzling victories, won mainly through the efforts of Thomas Young and Augustin Jean Fresnel (1788–1827). Between 1801 and 1803, Young made the memorable discoveries which provoked this revival of undulatory optics. The comparison of the aether that vibrates in a ray of light to the air that vibrates in a resonant tube led him to explain the alternately light and dark fringes that show in a place illuminated by two equal beams slightly inclined toward each other. The principle of interference, thus justified, allowed him to connect with the undulatory theory the explanation of the colors of thin laminae that Newton had demanded of the "fits of easy transmission and easy reflection" of the particles of light.

In 1815, Fresnel, who combined this principle of interference with the methods devised by Huygens, took up the theory of the phenomena of diffraction which had been discovered by Francesco Maria Grimaldi, S.J. (1618–1663), and that had remained a mystery to opticians. Fresnel's attempts at explaining these phenomena led him in 1818 to draw up a memoir which in a marked degree revealed the essential character of his genius—namely, a strange power of divination exercised independently of all rules of deductive reasoning. Despite the irregularity of his procedure, Fresnel made known very complicated formulae, the most minute details of which were verified by experiment, and long afterward justified according to the logical method of mathematicians. Never did a physicist conquer more important and more unthought truths, and yet never was there employed a method more capable of leading the common mind into error. Up to this time, the vibrations of aether in a ray of light had been supposed to be longitudinal, as it is in the air of a resonant tube, but in 1808, Etienne Louis Malus (1775–1812) discovered the polarization of light when reflected on glass; and, in 1817, when studying this phenomenon, Young was led to suppose that luminous vibrations are perpendicular to the ray which transmits them. Fresnel, who had conceived the same idea, completed an experiment (1816) in collaboration with Arago (1786–1853) which proved the view that luminous vibrations are transverse to the direction of propagation.

The hypothesis of transverse vibrations was, for Fresnel, the key to all the secrets of optics, and from the day he adopted it, he made discoveries with great rapidity. Among these discoveries were (a) the complete theory of the phenomena of polarization accompanying the reflection of light on the surface of contact of two isotropic media. The peculiarities which accompany total reflection of light on the surface of contact of two isotropic media. The peculiarities which accompany total reflection gave Fresnel an opportunity to display in a most striking manner his strange power of divination and thus to throw out a veritable challenge to logic. This divination was no less efficient in the second discovery: (b) In studying double refraction, Huygens limited himself to determining the direction of luminous rays in the interior of crystals now called *uniaxial* without, however, being able to account for the polarization of these rays; but with the aid of the wave-surface, Fresnel succeeded in giving the most elegant form to the law of the refraction of rays in biaxial crystals and in formulating rules by which rays polarize in the interior of all crystals, uniaxial as well as biaxial.

Although all these wonderful theories destroyed the theory of emission, the hypothesis of molecular attraction was far from losing ground. In fact, Fresnel thought he could find in the elasticity of the aether, which transmits luminous vibrations, the explanation of all the optical laws that he had verified by experiment, and he sought the explanation of this elasticity and its laws in the attraction which he believed to exist between the contiguous particles of this fluid. Being too little of a mathematician and too little of a mechanician to go very far in the analysis of such a problem, he left its solution to his successors. To this task, so clearly defined by Fresnel, Cauchy devoted the most powerful efforts of his genius as an algebraist, and thanks to this pupil of Laplace, the Newtonian physics of molecular attraction became an active factor in the propagation of the theory of undulatory optics.

Fresnel's discoveries did not please all Newtonians as much as they did Cauchy. Arago could never admit that luminous vibrations were transverse, notwithstanding that he had collaborated with Fresnel in performing the experiment by which this point was verified. Jean-Baptiste Biot (1774–1862), whose experimental researches were numerous and skillful and who had furnished recent optics with very valuable matter, remained strongly attached to the system of emission, by which he endeavored to explain all the phenomena that Fresnel had discovered and explained by the undulatory system. Moreover, Biot would not acknowledge himself defeated or regard the system of emission as condemned until Foucault (1819–1868) proved that light is propagated much more quickly in air than in water.

XXIX. Theories of Heat

The idea of the quantity of heat and the invention of the calorimeter intended for measuring the amount of heat emitted or absorbed by a body under given circumstances are due to Joseph Black (1728–1799) and Adair

Crawford (1749–1795), who, by joining calorimetry with thermometry, virtually created the science of heat, which science remained unborn as long as the only thing done was the comparison of temperatures. Like Descartes, Newton held that heat consisted in a very lively agitation of the smallest parts of which bodies are composed. By showing that a certain quantity of heat is furnished to ice which melts without, however, raising the temperature of the ice, that this heat remains in a "latent state" in the water resulting from the melting and that again becomes manifest when the water returns to ice, the experiments of Black and Crawford led physicists to change their opinion concerning the nature of heat. In it they beheld a certain fluid which combines with other matter when heat passes into the latent state and separates from it when heat is liberated again, and, in the new nomenclature that perpetuated the revolution brought about by Antoine-Laurent Lavoisier (1743–1794), this imponderable fluid was assigned a place among simple bodies and named *caloric*.

Air becomes heated when it is compressed and cools again when rarefied under the receiver of the pneumatic machine. Johann Heinrich Lambert (1728–1777), Horace de Saussure (1740–1779), and John Dalton (1766–1844) recognized the importance of this already old experiment, but it is to Laplace that we are indebted for a complete explanation of this phenomenon. The experiment proved to Laplace that at a given temperature, a mass of air contains a quantity of caloric proportional to its volume. If we admit the accuracy of the law of compressibility enunciated by Boyle and Mariotte, this quantity of heat, combined with a given mass of air also of a given temperature, is proportional to the volume of this air. In 1803, Laplace formulated these propositions in a short note inserted in Berthollet's *Statique chimique*. In order to verify the consequences which Laplace deduced therefrom concerning the expansion of gases, Louis-Joseph Gay-Lussac (1778–1850) began researches on this subject and, in 1807, on the variations of temperature produced when a gas contained in a receiver enters another receiver previously empty.

Laplace's views entail an evident corollary: To raise to a certain number of degrees the temperature of a gas of fixed volume, the communication of less heat is required than if this gas were expanded under an invariable pressure. Hence, a gas admits two distinct kinds of specific heat which depend on whether it is heated at constant volume or under constant pressure, the specific heat being greater in the latter case than in the former. Through these remarks, the study of the specific heat of gases was signalized as one of the most important in which experimenters could engage. The institute made this study the subject of a competition which called forth two notable memoirs, one by Delaroche and Bérard on the measurement of the specific heats of various gases under constant pressure, and the other by Desormes and Clément, published in 1812, on the determination of the increase of heat due to a given compression in a given mass of air. The experiments of Desormes and Clément enabled Laplace to deduce, in the case of air, the ratio of specific heat under constant pressure to specific heat under constant volume, and hence to test the ideas he had formed on the propagation of sound.

In applying to air the law of compressibility discovered by Boyle, Newton had attempted to calculate the velocity of the propagation of sound in this fluid, and the formula which he had established gave values very inferior to those furnished by experimental determination. Lagrange had already shown that by modifying Boyle's law of compressibility, this disagreement could be overcome; however, the modification was to be justified not by what Lagrange said but by what Laplace discovered. When sound is propagated in air by alternate condensations and rarefactions, the temperature at each point, instead of remaining unchanged, as Boyle's law supposed, is alternately raised and lowered about a mean value. Hence, velocity of sound was no longer expressed by the formula Newton had proposed; this expression had to be multiplied by the square root of the ratio of specific heat under constant pressure to specific heat under constant volume. Laplace had this thought in mind in 1803 (Berthollet, *Statique chimique*); its consequences were developed in 1807 by Poisson, his disciple. In 1816, Laplace published his new formula; fresh experiments by Desormes and Clément and analogous experiments by Gay-Lussac and Welter gave him tolerably exact values of the relation of the specific heats of gases. Henceforth the great geometer could compare the result given by his formula with that furnished by the direct determination of the velocity of sound—the latter, in meters per second, being represented by the number 340.889 and the former by the number 337.715. This agreement seemed a strong confirmation of the hypothesis of caloric and the theory of molecular action, to both of which it was attributable. It would appear that Laplace had a right to say (in 1822):

> The phenomena of the expansion of heat and vibration of gases lead back to the attractive and repellent forces sensible only at imperceptible distances. In my theory on capillary action, I have traced to similar forces the effects of capillarity. All terrestrial phenomena depend upon this species of force, just as celestial phenomena depend upon universal gravitation, and the study of these forces now seems to me the principal object of mathematical philosophy.

In 1824, a new truth was formulated from which was to be developed a doctrine which was to overturn, to a great extent, natural philosophy as conceived by Newton and Boscovich and carried out by Laplace and his disciples. Sadi Carnot (1796–1832), however, the author of this new truth, still assumed the correctness of the theory of caloric. He proposed to extend to heat engines the principle of the impossibility of perpetual motion recognized for engines of unchanging temperature, and he was led to the following conclusion: In order that a certain quantity of caloric may produce work of the kind that human industry requires, this caloric must pass from a hot to a cold body; when the quantity of caloric is given, as well as the temperatures to which these two bodies are raised, the useful work produced admits of a superior limit independent of the nature of the substances which transmit the caloric and of the device by means of which the transmission is effected. The moment that Carnot formulated this fertile truth, the foundations of the theory of caloric were shaken. In the hypothesis of caloric, however, how could the generation of heat by friction be explained? Two bodies rubbed together were found to be as rich in caloric as they had been; therefore, whence came the caloric evolved by friction?

As early as 1783, Lavoisier and Laplace were much troubled by the problem, which also arrested the attention of physicists, as in 1798, when Benjamin Thompson, Count Rumford (1753–1814), made accurate experiments on the heat evolved by friction, and, in 1799, when similar experiments were made by Sir Humphry Davy (1778–1829). In 1803, in addition to the notes in which Laplace announced some of the greatest conquests of the doctrine of caloric, Berthollet, in his *Statique chimique*, gave an account of Rumford's experiments, trying in vain to reconcile them with the prevailing opinion. Now these experiments, which were incompatible with the hypothesis that heat is a fluid contained in a quantity in each body, recalled to mind the supposition of Descartes and Newton, which claimed that heat was a very lively agitation of the small particles of bodies. It was in favor of this view that Rumford and Davy finally declared themselves.

In the last years of his life, Carnot consigned to paper a few notes which remained unpublished until 1878. In these notes, he rejected the theory of caloric as inconsistent with Rumford's experiments. "Heat," he added, "is therefore the result of motion. It is quite plain that it can be produced by the consumption of motive power and that it can produce this power. Wherever there is destruction of motive power there is, at the same time, production of heat in a quantity exactly proportional to the quantity of motive power destroyed; and inversely, wherever there is destruction of heat, there is production of motive power."

In 1842, Robert Mayer (1814–1878) found the principle of the equivalence between heat and work and showed that once the difference in two specific heats of a gas is known, it is possible to calculate the mechanical value of heat. This value differed little from that found by Carnot. Mayer's pleasing work exerted scarcely more influence on the progress of the theory of heat than did Carnot's unpublished notes. In 1843, however, James Prescott Joule (1818–1889) was the next to discover the principle of the equivalence between heat and work, and he conducted several of the experiments which, in his notes, Carnot had requested to have made. Joule's work communicated to the new theory a fresh impetus. In 1849, William Thomson, afterward Lord Kelvin (1824–1907), indicated the necessity of reconciling Carnot's principle with the thenceforth incontestable principle of the mechanical equivalent of heat; and in 1850, Rudolf Clausius (1822–1888) accomplished the task. Thus, the science of thermodynamics was founded. When, in 1847, Hermann von Helmholtz published his small work entitled *Über die Erhaltung der Kraft*, he showed that the principle of the mechanical equivalent of heat not only established a bond between mechanics and the theory of heat but also linked the studies of chemical reaction, electricity, and magnetism, and in this way physics was confronted with the carrying out of an entirely new program, whose results are at present too incomplete to be judged, even by scientists.[5]

5. The following are Duhem's references for this article: Roberto Almagia, "La Dottrina della marea nell' antichità classica et nel medio evo," *Memorie della Reale Accademia dei Lincei* (Rome, 1905). W. W. Rouse Ball, *An Essay on Newton's Principia* (London and New York, 1893). Idem, "Mémoires sur l'électrodynamique," *Collection de mémoires publiés par la société française de physique*, II–III (Paris, 1885–1887). Raffaello Caverni, *Storia del metodo sperimentale in Italia* (Florence, 1891–1898). Pierre Duhem, "Les Théories de la chaleur," *Revue des deux mondes* 129–130 (1895). Idem, *L'Évolution de la mecanique* (Paris, 1903). Idem, *Les Origines de la statique*, 2 vols. (Paris, 1905–1906). Idem, *Etudes sur Léonard de Vinci, ceux qu'il a lus et ceux qui l'ont lu*, 2 vols. (Paris, 1906–1909). Idem, *La Théorie physique, son objet et sa structure* (Paris, 1906). Idem, *SOZEIN TA PHAINOMENA. Essai sur la notion de théorie physique de Platon à Galilée* (Paris, 1908). Eugen Karl Duhring, *Kritische Geschichte der allgemeinen Principien der Mechanik* (2nd ed., Leipzig, 1877). Agost Heller, *Geschichte der Physik von Aristoteles bis auf die neueste Zeit*, 2 vols. (Stuttgart, 1882–1884). Gustav Hellmann, *Neudrucke von Schriften und Karten über Meteorologie und Erdmagnetismus*, 15 vols. (Berlin, 1893–1904). Emile Jouguet, *Lectures de mécanique, la mécanique enseignée par les auteurs originaux*, 2

vols. (Paris, 1908–1909). Hermann Klein, *Die Principien der Mechanik, Historisch und Kritisch Dargestellt* (Leipzig, 1872). Kurd Lasswitz, *Geschichte der Atomistik vom Mittelalter bis Newton*, 2 vols. (Hamburg and Leipzig, 1890). Guillaume Libri, *Histoire des sciences mathématiques en Italie, depuis la renaissance des lettres jusqu'à la fin du XVII siècle*, 4 vols. (Paris, 1838–1841). Ernst Mach, *Die Mechanik in ihrer Entwickelung, historisch-kritisch dargestellt* (6th ed., Leipzig, 1908). Blaise Pascal, *Oeuvres*, eds. Brunschvicg and Boutroux, 3 vols. (Paris, 1908). Pierre Sue, *Histoire du galvanisme et analyse des différents ouvrages publiés sur cette découverte, depuis son origine jusqu'à nos jours*, 4 vols. (Paris, 1802–1803). Julien Thirion, "Pascal, l'horreur du vide et la pression atmosphérique," *Revue des questions scientifiques*, 3rd series, 12–14 (1907–1909). Charles Thurot, "Recherches historiques sur le principe d'Archimede," *Revue archéologigue* (1868–1869), 18–20. Isaac Todhunter, *A History of Mathematical Theories of Attraction and the Figure of the Earth from Time of Newton to That of Laplace*, 2 vols. (London, 1873). Isaac Todhunter and Karl Pearson, *A History of the Theory of Elasticity*, 2 vols. (Cambridge, 1886–1893). Giovanni Battista Venturi, *Commentari sopra la storica e le teorie dell' Ottica* (Bologna, 1814). Augustin Jean Verdet, *Introduction aux oeuvres d'Augustin Fresnel*, I (Paris, 1866–1870), pp. ix–xcix. Gustav Heinrich Wiedemann, *Die Lehre von der Elektricität*, 3 vols. (2nd ed., Brunswick, 1893–1895). Emil Wohlwill, "Die Entdeckung des Beharrungsgesetzes," *Zeitschrift für Volkerpsychologie und Sprachwissenschaft* (1883–1884), 14–15. Idem, *Galilei und sein Kampf fur die Copernicanische Lehre* (Hamburg and Leipzig, 1909).

9

The Nature of
Mathematical Reasoning[1]

This article, published the year Henri Poincaré, the great French mathematician and philosopher of science and mathematics, died, is a critique of the first chapter of Poincaré's magnum opus, La Science et l'hypothèse. *Its interest lies not only in Duhem's critique of so significant a work but also in the degree to which Duhem does or does not historicize the study of mathematical and deductive reasoning, as he had done with physical theory and induction.*

I

Until now, we have been accustomed to regard mathematical reasoning as the most perfect known model of deductive reasoning; we believed that no demonstration is received by arithmeticians or geometers as absolutely rigorous or convincing unless it is reducible to a syllogism or to a terminating series containing a limited number of syllogisms.

Some years ago, in an article that created a sensation, Henri Poincaré asserted that this commonly received opinion was erroneous.[2] According to Poincaré, arithmetic frequently uses a reasoning which is not equivalent to a series of syllogisms of *limited number*; in reality, it condenses an infinity of successive syllogisms.

The use of this mode of reasoning characterizes mathematical demon-

---

1. ["La nature du raisonnement mathématique," *Revue de philosophie* 12 (1912): 531–543.]

2. H. Poincaré, *La Science et l'hypothèse*, part 1, chap. 1: "Sur la nature du raisonnement mathématique." [*Science and Hypothesis* (New York: Dover, 1952).]

stration and distinguishes it from simple syllogistic deduction; at the same time it explains how mathematical demonstration enjoys a creative fertility that simple syllogistic deduction cannot possess.

In the past, the mode of reasoning considered by Poincaré was often called *process of recurrence* (*procédé par récurrence*). Today, in order to affirm that it cannot be reduced to simple syllogistic deduction, it is called *complete induction* (*induction complète*). Modern logicians have devoted all their attention to this complete induction.

Let us recall what this reasoning by recurrence or complete induction is like.

First, we *verify* that a certain arithmetical proposition, which we call proposition P, is true for the number 1.

Second, we *demonstrate* by a purely syllogistic path that if it is true for any whole number n whatever, it is true for the successive whole number $(n + 1)$.

Thus, we *conclude* that proposition P is true for all whole numbers.

Poincaré writes,

> The essential characteristic of reasoning by recurrence is that it contains, condensed, so to speak, in a single formula, an infinite number of syllogisms. We shall see this more clearly if we enunciate the syllogisms one after another. They follow one another, if one may use the expression, in a cascade. The following are the hypothetical syllogisms:—The theorem is true of the number 1. Now, if it is true of 1, it is true of 2. Now, if it is true of 2, it is true of 3; hence it is true of 3, and so on. We see that the conclusion of each syllogism serves as the minor of its successor. Further, the majors of all our syllogisms may be reduced to a single form. If the theorem is true of $n - 1$, it is true of n.
>
> We see, then, that in reasoning by recurrence we confine ourselves to the enunciation of the minor of the first syllogism, and the general formula which contains as particular case all the majors. This unending series of syllogisms is thus reduced to a phrase of a few lines. . . .
>
> I asked at the outset why we cannot conceive of a mind powerful enough to see at a glance the whole body of mathematical truth. The answer is now easy. A chess player can combine for four or five moves ahead; but, however extraordinary a player he may be, he cannot prepare for more than a finite number of moves. If he applies his faculty to Arithmetic, he cannot conceive its general truths by direct intuition alone; to prove even the smallest theorem he must use reasoning by recurrence, for that is the only instrument which enables us to pass from the finite to the infinite.[3]

3. Poincaré, *La Science et l'hypothèse*, pp. 20–22. [*Science and Hypothesis*, pp. 9–11.]

II

The ancients did not use the mode of reasoning we have just spoken of. It seems to have been introduced for the first time in mathematics by Francesco Maurolico (*Maurolycus*). Maurolycus used this process in his *Arithmeticorum libri duo*, written in 1557 and published in Venice in 1575. He emphasizes its novelty in the preface.

If reasoning by recurrence were truly what characterizes mathematical demonstration, it would be surprising that mathematicians had waited so long to have recourse to it. Complete induction cannot be the sole source of mathematical fertility, given that Euclid, Archimedes, Apollonius, and Cardano had no need for it in order to produce their discoveries.

Further, the process of recurrence has absolutely nothing that essentially distinguishes it from the other forms of deductive reasoning. If one believes that it differs essentially from—and that it cannot in any way be reduced to—a series of syllogisms, that is because it is ordinarily displayed in abbreviated form; we overlook a complement which is easy to establish but which is required by absolute rigor.

Doubtless, as soon as we have these two truths:

Proposition P is true for the number 1;
If proposition P is true for the number n, it is true for the number $(n + 1)$.

we begin to unroll in our minds the series of "syllogisms in a cascade" described by Poincaré; and we immediately recognize that nothing is capable of stopping that sorites, that it can be pursued up to whatever number we wish to reach, and we conclude directly that proposition P is true of all whole numbers.

> Why, then, is this view imposed upon us with such an irresistible weight of evidence? It is because it is only the affirmation of the power of the mind which knows it can conceive of the indefinite repetition of the same act, when the act is once possible. The mind has a direct intuition of this power, and experiment can only be for it an opportunity of using it, and thereby of becoming conscious of it.[4]

We agree with Poincaré that, in this way, our reason is persuaded of the generality of proposition P even before a reasoning, fashioned from a limited number of syllogisms, has demonstrated this generality. But if the cer-

---

4. Poincaré, *La Science et l'hypothèse*, pp. 23–24. [*Science and Hypothesis*, p. 13.]

tainty that we have of this generality is absolute, it is because we know that we can convict of a contradiction anyone who would contest the generality of this proposition.

Let us show this reduction to absurdity in detail.

III

We assume that the following two lemmas have been demonstrated:

> LEMMA I: *Proposition P is made true by the number 1*
> LEMMA II: *If proposition P is true for whole number* n, *it is still true for the immediately succeeding whole number* (n + 1).

We affirm that *proposition* P *is true for all whole numbers.*

Let us assume, in fact, that it is false for a certain whole number, and let p be this number; in virtue of lemma I, this number p is certainly greater than 1, such that $(p - 1)$ is a whole number.

We can imagine that two cases present themselves:

> *Case 1*: Proposition P is true for all whole numbers from 1 to $(p - 1)$.
> *Case 2*: Proposition P is false for one or more of the numbers of which 2 is the smallest and $(p - 2)$ is the greatest.

In case 1, proposition P would be true for whole number $(p - 1)$ and false for whole number p; according to lemma II, this is impossible.

In case 2, let us enumerate, in order of increasing magnitude, the series of $(p - 2)$ whole numbers from 2 to $(p - 1)$. We encounter, one after another, all the numbers of this series for which proposition P is false. Let q be the one we encounter first. According to lemma I, q is necessarily greater than 1, such that $(q - 1)$ is a whole number. Proposition P will therefore be true for number $(q - 1)$ and false for q. According to lemma II, this is impossible.

The generality of proposition P is thus established by a reasoning that requires only a finite number of syllogisms.

IV

We could have substituted, for the reasoning by absurdity we have just used, the following demonstration, which differs very little from it:

If proposition P is not true for all whole numbers, let us consider the infinite or finite set of whole numbers for which it is not true. In that set, there is a whole number p which is smaller than all the others. In virtue of lemma I, this number p is necessarily greater than 1, such that $(p-1)$ is a whole number. Thence, proposition P, true for number $(p-1)$, would be false for number p—which, by lemma II, would be impossible.

This demonstration clearly admits the validity of the following THEO-REM: *In any finite or infinite set of different whole numbers, there is one smaller than all the others.*

It is assuredly to this demonstration that Poincaré alluded when he wrote:

> The views upon which reasoning by recurrence is based may be exhibited in other forms; we may say, for instance, that in any finite collection of different integers there is always one which is smaller than any other. We may readily pass from one enunciation to another, and thus give ourselves the illusion of having proved that our reasoning by recurrence is legitimate. But we shall always be brought to a full stop—we shall always come to an indemonstrable axiom, which will at bottom be but the proposition we had to prove translated into another language. We cannot therefore escape the conclusion that the rule of reasoning by recurrence is irreducible to the principle of contradiction.[5]

According to us, the proposition—*In any finite or infinite set of different whole numbers, there is one smaller than all the others*—is not an indemonstrable AXIOM, but a theorem that can be demonstrated with the help of a finite number of syllogisms, in the following way:

Let E be the said set and p a number which is an element of this set. There can be three cases.

Case 1: p is equal to 1.

In that case, 1 is the smallest number of set E.

Case 2: p is greater than 1, such that $(p-1)$ is a whole number; among the $(p-1)$ first whole numbers, none is an element of set E.

In that case, p is the smallest number of set E.

Case 3: p is greater than 1; among the whole numbers that go from 1 to $(p-1)$, there is one or more that belong to set E.

5. Poincaré, *La Science et l'hypothèse*, pp. 22–23. [*Science and Hypothesis*, p. 12.]

In that case, let us enumerate in thought, in order of increasing magnitude, the limited series of numbers that go from 1 to $(p - 1)$; we will encounter, one after another all the numbers of this series that are elements of set E; the one we will encounter before all the others will be the smallest number of set E.

In all cases, then, there is a whole number smaller than all the others in set E.

V

Perhaps someone can still object that the preceding demonstrations imply a postulate that cannot be justified by means of a limited number of syllogisms; the postulate would be the following:

Given any whole number whatever n, *the operation that consists in enumerating in thought, in the order of increasing magnitude, the series of whole numbers from* 1 *to* n (*or in counting mentally from* 1 *to* n) *is a possible operation.*

That proposition is neither postulate nor axiom; it is a simple truism.

In fact, all whole numbers are defined as the result of the addition of unity to the preceding whole number; thus, the definition of any whole number implies enumeration, in order of increasing magnitude, of all the whole numbers preceding it. Moreover, when I say *n is a whole number*, it is the same as if I said *we can count to n*.

Let us illustrate this conclusion with an example.

Let us imagine that I pronounce these words: *assume the whole number represented by unity followed by ten zeros.* These words have no meaning except in virtue of the following proposition: *it is possible to count until one reaches a number represented, in decimal enumeration, by unity followed by ten zeros.* If I did not admit the correctness of that proposition, I could truly write unity followed by ten zeroes, but I could not say that this sign represents a whole number.

We therefore are right to consider this assertion as a truism: *We cannot conceive a whole number so great that we cannot count up to that number,* since to conceive that something is a whole number is to conceive that one could reach that thing by counting.

VI

If we are careful to completely exhibit reasoning by recurrence, and not to omit the argument by reduction to absurdity, which must complete it, we recognize that it is reduced to a series of a finite number of syllogisms.

It is true that, in general, the explicit presentation of this complement of reasoning by recurrence is neglected in arithmetical or algebraic treatises. If it is left out in this way, no doubt it is because it is judged too easy to imagine for its development to be useful. It is no less true that this complement is indispensable if one wishes to render reasoning by recurrence into an absolutely rigorous process of demonstration.

Something happens here that is similar to what takes place in a number of demonstrations that achieve a limit. A first part of the demonstration gives the mind a direct intuition of the truth we wish to prove; but in order to transform this intuition into absolute certainty, acquired by rigorous deduction, we must have recourse to a demonstration by absurdity.

We know what care the ancients, and Archimedes in particular, brought to this second part of the demonstration; if they ever needed to use a process by recurrence, one can believe that they would not have neglected to expose the complementary by reduction to absurdity; moderns, because of a desire for brevity, have passed over it in silence.

VII

Reasoning by recurrence does not require recourse to that infinite series of syllogisms "in cascade" that Poincaré considered. Therefore, the properties of mathematical propositions that the noted algebraist attempted to explain by means of that unlimited series of syllogisms must be explained in another way.

Now, Poincaré saw in this unlimited series of syllogisms the reason for the inexhaustible fertility of mathematics; he also found in it the cause that conferred generality to arithmetical propositions. He said on the subject of complete induction:

> This instrument is always useful, for it enables us to leap over as many stages as we wish; it frees us from the necessity of long, tedious, and monotonous verifications which would rapidly become impracticable. Then when we take in hand the general theorem it becomes indispensable, for otherwise we should ever be approaching the analytical verification without ever actually reaching it.[6]

If reasoning by recurrence furnishes us a general conclusion, if it allows us to affirm that *proposition P is true for all whole numbers*, we cannot, as we

6. Poincaré, *La Science et l'hypothèse*, p. 22. [*Science and Hypothesis*, p. 11.]

have said, explain it in the same way as Poincaré did. From where does the generality of that proposition arise? Evidently, from the fact that lemma II, *If proposition P is true for number* n, *it will also be true for the whole number* (*n* = 1), is not applied only to such or such particular number, but to whatever whole number one puts in the place of letter *n*. And if this lemma presents such a character, it owes it to the fact that the reasoning by which it has been obtained did not bear on this or that particular number but on the general notion of whole number. The abstraction that allows us to conceive this general notion of whole number is here, as in all circumstances, the single cause of the generality noted in some of our judgments.

VIII

Let us now see how the fertility of mathematical sciences should be explained. The difficulty that had first captured Poincaré's attention lies there.

The very possibility of mathematical science seems an insoluble contradiction. If this science is only deductive in appearance, from whence is derived that perfect rigor which is challenged by none? If, on the contrary, all the propositions which it enunciates may be derived in order by the rules of formal logic, how is it that mathematics is not reduced to a gigantic tautology? The syllogism can teach us nothing essentially new, and if everything must spring from the principle of identity, then everything should be capable of being reduced to that principle. Are we then to admit that the enunciations of all the theorems with which so many volumes are filled are only indirect ways of saying that A is A?

No doubt we may refer back to axioms which are at the source of all these reasonings. If it is felt that they cannot be reduced to the principle of contradictions, if we declined to see in them any more experimental facts which have no part or lot in mathematical necessity, there is still one resource left to us: we may class them among *a priori* synthetic views. But this is no solution of the difficulty—it is merely giving it a name; and even if the nature of the synthetic views had no longer a mystery, the contradiction would not have disappeared; it would have only been shirked. Syllogistic reasoning remains incapable of adding anything to the data that are given it; the data are reduced to axioms, and that is all we should find in the conclusions.

No theorem can be new unless a new axiom intervenes in its demonstration; reasoning can only give us immediately evident truths borrowed from direct intuition; it would only be an intermediary parasite. Should we not therefore have reason for asking if the syllogistic apparatus serves only to disguise what we have borrowed?

The contradiction will strike us the more if we open any book on mathematics; on every page the author announces his intention of generalizing some proposition already known. Does the mathematical method proceed from the particular to the general, and, if so, how can it be called deductive?

Finally, if the science of number were merely analytical, or could be analytically derived from a few synthetic intuitions, it seems that a sufficiently powerful mind could with a single glance perceive all its truths; nay, one might even hope that someday a language would be invented simple enough for these truths to be made evident to any person of ordinary intelligence.

Even if these consequences are challenged, it must be granted that mathematical reasoning has of itself a kind of creative virtue, and is therefore to be distinguished from the syllogism.[7]

We believe that all these difficulties noted by Poincaré vanish if one makes the following remark:

Mathematics is not derived wholly by syllogistic means from axioms alone but from axioms and *definitions*. The syllogisms with which any chapter of mathematical science begins have axioms or else theorems demonstrated in preceding chapters for major premises; but the propositions that serve as minor premises are the definitions of mathematical notions one proposes to treat in that chapter.

For example, the syllogisms one finds at the beginning of the theory of fractional numbers have axioms or else theorems demonstrated in the theory of whole numbers for major premises; but their minor premises are propositions taken among those that serve to define fractional numbers. And it is because of this that the conclusions state properties of fractional numbers.

Now, in truth, axioms that bear on any part of mathematical sciences are of limited number, and even in small enough number. They can all be stated explicitly, and one tries to give these explicit statements as completely as possible. If mathematics were derived by syllogistic means from these axioms alone, it would enclose nothing more than that which would be contained virtually in these simple and few propositions, and we would be correctly surprised that they would feel the need to produce an unlimited series of more and more complicate theorems.

But mathematics is not virtually contained in axioms alone; it is the result produced by the application of axioms to definitions. Now, though the axioms of a mathematical science are of limited number—and it is impossible for us to increase this number, since it is impossible for us to

---

7. Poincaré, *La Science et l'hypothèse*, pp. 9–11. [*Science and Hypothesis*, pp. 1–3.]

order the truth of a new axiom—the definitions themselves form an unlimited multitude; it is always possible for us to add to the previously defined and studied mathematical concepts new mathematical notions obtained by combination, modification, and generalization of the previously defined and studied notions. In this direction, nothing comes to terminate the power that our intellect has of composing new ideas.

It is therefore to the definitions, and not to the axioms, that mathematics owes the power that resides in it of developing an unlimited series of truly and always new theorems; it is through definitions, and not through axioms, that the creative activity of our intellect manifests itself in mathematics.

It is now easy to understand what one means by generalization in mathematics.

We have demonstrated a proposition A, which states a certain property of a mathematical notion *a*. We compose a mathematical notion *b*, which includes notion *a* as particular case. Finally, with respect to this notion *b*, we demonstrate a proposition B that restores proposition A when we substitute for notion *b* its particular determination *a*. Theorem B is a generalization of theorem A.

For example, we have demonstrated a certain property of whole numbers. We create the notion of fractional number, which encloses, in particular, the notion of whole number. We finally demonstrate a certain property of fractional number. This property is such that if we reduce this number to being only a whole, it will become identical to the property of a whole number that we had first demonstrated. The theorem with respect to fractional number is a generalization of the theorem with respect to whole number.

We see clearly that that which renders the generalization of mathematical theorems possible is the generalization of definitions.

We think we have sufficiently established, in the preceding, that mathematical demonstration is pursued syllogistically in the same way as any other deductive science. What distinguished it from other deductive sciences is not the form of the reasoning it uses; it is the nature of the notions and propositions to which it applies this reasoning.

10

Logical Examination of Physical Theory[1]

In May 1913, Duhem wrote an account of his works as a submission toward his membership to the Académie des Sciences. The report was a lengthy essay divided into three parts, the first concerning his scientific, the second his philosophical, and the third his historical works. This is the second part, a significant essay not only because it is Duhem's overview of his philosophy of science toward the end of his life but also because he places his philosophy of science within the context of other philosophies at the time: Positivism, Pragmatism, Inductivism, and so on.

Theoretical physics may be treated in the fashion of Cartesians and atomists. They resolve the bodies perceived by the senses and instruments into immensely numerous and much smaller bodies of which reason alone has knowledge. Observable motions are regarded as the combined effects of the imperceptible motions of these little bodies. These little bodies are assigned shapes which are few in number and well defined. Their motions are given by very simple and entirely general laws. These bodies and these motions are, strictly speaking, the only real bodies and the only real motions. When they have been suitably combined and recognized as being together capable of producing effects equivalent to the phenomena we

1. ["Examen logique de la théorie physique," *Revue scientifique* 51 (1913): 737–740. Same as "Notice sur les titres et travaux scientifiques de Pierre Duhem, rédigée par lui-même lors de sa candidature à l'académie des sciences (mai 1913)," *Mémoires de la société des sciences physiques et naturelles de Bordeaux*, ser. 7, vol. 1 (1917), part 2.]

observe, it is claimed that the explanation of these phenomena has been discovered.

Our own view, Energetics, does not proceed in this manner. The principles it embodies and from which it derives conclusions do not aspire at all to resolve the bodies we perceive or the motions we report into imperceptible bodies or hidden motions. Energetics presents no revelations on the true nature of matter. Energetics claims to explain nothing. Energetics simply gives general rules of which the laws observed by the experimentalist are particular cases.

Alternatively, theoretical physics may be conceived in the manner of Newtonians. They reject all hypotheses about imperceptible bodies and hidden motions, of which the bodies and motions accessible to the senses and instruments may be composed. The only principles admitted are very general laws known through induction, based on the observation of facts.

Energetics does not follow the method of the Newtonians. Energetics recognizes without doubt an experimental origin to the principles it admits, in the sense that observation has suggested them and experiment has many times counseled their modification. But Energetics does not regard these experiments, which explain the possible genesis of the principles that Energetics embodies, as capable of conferring any certainty whatever on these principles. Energetics regards these principles as pure postulates, or arbitrary decrees of reason. When they produce numerous consequences conforming to experimental laws, Energetics regards them as playing their assigned roles well. Agreement with the teaching of observation is not, therefore, as the Newtonian method would require, the beginning of physical theory; it has its place at the end.

Is Energetics being wise when it refuses equally to follow the method of Cartesians and atomists and the method of the Newtonians? Does careful examination of the epistemological methods of physics justify the attitude that Energetics adopts? To this question we have replied: Yes.

We have criticized the method of the Cartesians and atomists for not being autonomous.[2] The physicist who wishes to follow it cannot use exclusively the methods proper to physics since, behind perceptible bodies and motions which he regards as appearances, he aspires to get hold of other bodies and other appearances, which are the only true ones. Here he enters the domain of cosmology. He no longer has the right to shut his ears

---

2. "Quelques réflexions au sujet des théories physiques," *Revue des questions scientifiques* 1 (1892) [chapter 1 of this volume]; *La Théorie physique, son objet et sa structure* (Paris, 1906).

to what metaphysics wishes to tell him about the real nature of matter; hence, as a consequence, through dependence on metaphysical cosmology, his physics suffers from all the uncertainties and vicissitudes of that doctrine. Theories constructed by the method of the Cartesians and atomists are also condemned to infinite multiplication and to perpetual reformulation. They do not appear to be in any state to assure consensus and continual progress to science.

We have criticized the Newtonian method for being impractical.[3]

A science may progress following the Newtonian method while its epistemological methods remain those of common sense. When science no longer observes facts directly but substitutes for them measurements, given by instruments, of magnitudes that mathematical theory alone defines, induction can no longer be practiced in the manner that the Newtonian method requires.

> An experiment in physics is not simply the observation of a phenomenon[. . . .]
> An experiment in physics is the precise observation of a group of phenomena, accompanied by the interpretation of these phenomena. This interpretation replaces the concrete data really gathered by observation with abstract and symbolic representations that correspond to them by virtue of physical theories accepted by the observer.[4]

From this truism follow numerous consequences strongly opposed to the idea of a science in which each principle may be supplied by induction:

> Physicists can never submit an isolated hypothesis to the control of experiment, but only a whole group of hypotheses. When an experiment is in disagreement with their predictions, it tells them that at least one of the hypotheses which constitute this group is erroneous and must be modified, but it does not tell them which one must be changed.[5]

3. "Quelques réflexions au sujet de la physique expérimentale," *Revue des questions scientifiques* 3 (1894) [chapter 4 of this volume]; *La Théorie physique, son objet et sa structure*.

4. "Quelques réflexions au sujet de la physique expérimentale," *Revue des questions scientifiques* 3 (1894) [chapter 4 of this volume]; *La Théorie physique, son objet et sa structure*.

5. "Quelques réflexions au sujet de la physique expérimentale," *Revue des questions scientifiques* 3 (1894) [chapter 4 of this volume]; *La Théorie physique, son objet et sa structure*.

Here we are a long way from the mechanism of experiment such as people who are strangers to its functioning readily imagine it. One commonly thinks that each of the hypotheses used by physics may be taken in isolation, submitted to the control of experience, and then, when varied and repeated proofs have established its value, placed into the totality of science in an almost definitive fashion. In reality, it is not so; physics is not a machine that lets itself be disassembled. We cannot address each piece in isolation and wait to adjust it until its soundness has been minutely controlled. Physical science is an organism one must take hold of in one piece. It is an organism in which one part cannot be made to function without the parts most distant from it coming into play, some more, some less, all to some degree. If some difficulty, some *malaise*, reveals itself in its functioning, the physicist will be obliged to discover the organ that needs to be adjusted or modified without it being possible for him to isolate that organ and examine it on its own. The clockmaker to whom one gives a clock that does not work takes all the wheels out of it and examines them one by one until he finds the bent or broken one. But the doctor to whom one brings a sick person cannot dissect the patient to establish his diagnosis; he must discover the seat of the illness only through the inspection of effects produced on the whole body. The physicist responsible for repairing a rickety theory resembles the latter, not the former.

Physical theory is not an explanation of the inorganic world; still less is it an inductive generalization of the teachings of experience. So what is it?[6] Is theory simply, as the pragmatists would like it, a tool that gives us truths of empirical knowledge in the easiest manner, permits us to make faster and more profitable use of it in our action on the external world, but does not teach us anything about this world that we would not already have been taught by experience alone?

Or, on the contrary, does theory teach us about what is real—something that experience has not taught us and would not be able to teach us, something that would be transcendent to purely empirical knowledge?

If we were to respond affirmatively to this last question, we would be saying that physical theory is *true*, that it has value as *knowledge*. If, on the

6. "Physique et métaphysique," *Revue des questions scientifiques* 2 (1893) [chapter 2 of this volume]; *La Théorie physique, son objet et sa structure.* "Sur un fragment, inconnu jusqu'ici, de l' *Opus tertium* de Roger Bacon," *Archivium Franciscannum historicum* 1 (1908). "La valeur de la théorie physique, à propos d'un livre récent," *Revue générale des sciences pures et appliquées* (1908) [appendix to *Aim and Structure of Physical Theory*, pp. 312–335].

contrary, it is the first question that constrains us to say yes, we would have to say also that physical theory is not *true* but simply *convenient*; that it has no value as knowledge but solely *practical value*.

> When the physicist, turning his attention to the science he is constructing, submits the procedures that he has used to a rigorous examination, he discovers nothing able to introduce into the edifice the least particle of truth, except experimental observation. Of propositions attempting to state the facts of experience and of these alone we may say: *It is true*: or: *It is false*. Of these alone we may assert that they will not permit illogicality, and that of two contradictory propositions one at least must be rejected. As for propositions introduced by theory, they are neither *true* nor *false*. They are simply *convenient* or *inconvenient*. If the physicist finds it convenient to construct two chapters of physics with the aid of hypotheses that contradict each other, he is free to do so. The principle of contradiction is able to judge truth and falsity decisively. It has no ability to decide what is useful and what is not. Therefore, to require physical theory to observe a rigorous logical unity in its development would be to exert an unjust and insupportable tyranny on the intellect of the physicist.
>
> When, after having submitted the science that concerns him to this minute examination, the physicist returns to his own concerns, when he takes notice of the tendencies that direct the steps of his reasoning, he recognizes at the same time that all his most profound and most powerful aspirations are crushed by the heartbreaking conclusions of his analysis. No, he cannot bring himself to see in physical theory only a collection of practical procedures, a bag full of tools. No, he cannot believe that physical theory only catalogs knowledge accumulated through empirical science, without changing the nature of this knowledge in the least, and without imprinting it with a character that experience alone would not be able to engrave at all. If there were no more in physical theory than critical examination had shown him in it, he would stop devoting his time and his efforts to a work of so little importance. *The study of the method of physical science is powerless to show the physicist the reason that leads him to construct physical theory.*
>
> No physicist, however positivistic we imagine him to be, would be able to deny this declaration. But his positivism must be sufficiently rigorous that he would not go beyond this declaration, and say that his efforts towards a physical theory, which is always more unitary and always more general, are reasonable, although critical examination of the method of physical science has not been able to discover a reasonable basis for it. Such a basis might be expressed precisely in the following propositions:
>
> Physical theory gives us a type of knowledge of the external world not reducible to purely empirical knowledge. This knowledge comes neither from experience nor from the mathematical procedures the theory employs. Purely logical dissection of the theory would not discover the crack by which this

knowledge introduces itself into the edifice of physics, through a route which the physicist can no more deny is real, any more than he can describe its course. This knowledge derives from a truth other than the truths which our instruments are appropriate to grasp. The order into which theory places the results of observation does not find its full and complete justification in its practical or aesthetic aspects. We come to see, on the other hand, that this order is, or tends to become, a *natural classification*. Through an analogy the nature of which escapes the grasp of physics, but the existence of which imposes itself on the mind of the physicist as certain, we come to know that this order corresponds better and better to a certain overarching order.

In a word, the physicist is forced to recognize that *it would be irrational to work towards the progress of physical theory if that theory were not the more and more clear, and more and more precise reflection of a metaphysics. The belief in an order transcending physics is the sole reason for the existence of physical theory.*

The attitude, hostile or favorable by turns, which all physicists take towards this declaration is captured in this saying of Pascal: "Our powerlessness to prove anything is invulnerable to all Dogmatism; our idea of truth is invulnerable to all Skepticism [*Pyrrhonisme*]."[7]

Separated from the various schools of pragmatists on the subject of the value of physical theory, we do not take our stand, in any circumstance, among the number of their followers. The analysis we have given of experiments in physics shows fact to be completely interpenetrated by theoretical interpretation, to the point that it becomes impossible to express fact, in isolation from theory, in such experiments. This analysis has found great favor on the side of many pragmatists. They have applied it to the most diverse fields: to history, to exegesis, to theology. We do not deny that this extension is legitimate *to some extent*. However different the problems may be, it is always the same human intellect that exerts itself to resolve them. In the same way, there is always something common in the several procedures reason employs. But if it is good to notice the analogies between our diverse scientific methods, it is on condition that we do not forget the differences separating them. And when we compare the method of physics, so strangely specialized in the application of mathematical theory and by the use of instruments of measurement, to other methods, there are surely more differences to describe than analogies to discover.

We accept that physical theory is able to obtain a certain type of knowledge of the nature of things; but this knowledge, which is purely analogi-

7. "La valeur de la théorie physique, à propos d'un livre récent" [appendix to *Aim and Structure of Physical Theory*].

cal, appears to us as the terminus of theoretical progress, as the limit which theory endlessly approaches without ever reaching it. On the contrary, the schools of the Cartesians and atomists place hypothetical knowledge of the nature of things at the origin of physical theory. If, therefore, we separate ourselves from the pragmatists, it is not to take a place among the Cartesians or the atomists.

The school of the neo–atomists, the doctrines of which center on the concept of the electron, have taken up again with supreme confidence the method we refuse to follow. This school thinks its hypotheses attain at last the inner structure of matter, that they make us see the elements as if some extraordinary ultra-microscope were to enlarge them until they were made perceptible to us.

We do not share this confidence. We are not able to recognize in these hypotheses a clairvoyant vision of what there is beyond sensible things; we regard them only as *models*. We have never denied the usefulness of these models, dear to physicists of the English school.[8] We believe they lend an indispensable aid to minds more broad than deep, more able to imagine the concrete than to conceive the abstract. But the time will undoubtedly come when, through their increasing complications, these representations or models will cease to be aids for the physicist. He will regard them instead as embarrassments and impediments. Putting aside these hypothetical mechanisms, he will carefully release from them the experimental laws they have helped to discover. Without pretending to explain these laws, he will seek to classify them according to the method we have just analyzed to understand them within a modified and a broader Energetics.

8. "L'Ecole anglaise et les théories physiques," *Revue des questions scientifiques* 2 (1893) [chapter 3 of this volume]; *La Théorie physique, son objet et sa structure.*

11

Research on the History of Physical Theories[1]

This, the third part of Duhem's report to the Académie des Sciences, gives an overview of his own historical works. The philosophical interest of this piece lies in seeing it as an application of Duhem's historiographical concerns and as an illustration of Duhem's theses about physical theory and its relation to metaphysics.

All abstract thought requires the control of facts; all scientific theories call for comparison with experience. Our logical considerations about the proper method of physics cannot be judged rationally unless they are confronted with the teachings of history. We must now apply ourselves toward gathering these teachings.

During antiquity, the Middle Ages, and the Renaissance, there was hardly more than one part of physical theory in which mathematical theory had sufficient development and observation had sufficient precision for us to discuss their mutual relations; this part is astronomy.

With regard to the nature and value of astronomical theory, one might say that the Greek mind, so admirably supple, penetrating, and varied, conceived all the systems that our time has seen flourish again.[2] But

1. [Recherches sur l'histoire des théories physiques, Part III of "Notice sur les titres et travaux scientifiques de Pierre Duhem, rédigée par lui-même lors de sa candidature à l'académie des sciences (mai 1913)," *Mémoires de la société des sciences physiques et naturelles de Bordeaux*, ser. 7, vol. 1 (1917).]

2. *SOZEIN TA PHAINOMENA*—*Essai sur la notion de théorie physique de Platon à Galilée* (Paris, 1908).

among these systems, there is one that wins the approbation of the most profound thinkers. It can be summarized in the following principle that Plato taught to those who wanted to work in astronomy: "When taking certain assumptions as our point of departure, one must attempt to save what appears to the senses—*Tinon upotethenton,*[. . .] *sozein ta phainomena.*" And this principle spans the Arabic, Jewish, and Christian Middle Ages, is repeated at the time of the Renaissance, is explained, specified, or contested, up to the day when Andreas Osiander formulates it thus, in the preface that he placed at the head of Copernicus's book: *"Neque enim necesse est eas hypotheses esse veras, imo, ne verisimiles quidem, sed sufficit hoc unum si calculum observationibus congruentem exhibeant."*[3] For two thousand years, therefore, the majority of those who reflected on the nature and value of the mathematical theory used by the physicists agreed to proclaim the axiom that Energetics came to take as its own: The first postulates of physical theory are not given as affirmations of certain suprasensible realities; they are general rules which would have played their role admirably if the particular consequences deduced from them agreed with the observed phenomena.

The method followed by Energetics is not an innovation; it can call forth the most ancient, most continuous, and most noble tradition for itself. But what should we say about the essential notions and fundamental principles of that science? Logic does not require any justification of Energetics when it defines these notions and posits these principles; logic leaves it free to posit its foundations as it wishes, as long as, having reached its zenith, the edifice is capable of accommodating without constraint or disorder the laws ascertained by the experimenter. Is that to say that Energetics defines these notions haphazardly and posits these principles without reason? Not at all. Although logic does not impose any constraint upon Energetics, the teachings of history are an extremely sure and meticulous guide for it; the remembrance of past attempts, and of their happy or unhappy fate, prevents Energetics from receiving hypotheses which have led older theories to their ruin, or persuades it to adopt ideas which have already been shown to be fruitful. Energetics would not be able to prove its postulates, and does not have to prove them, but by retracing the vicissitudes they have gone through before they came to have their present form, it can gain our confidence for them—that is, it can obtain some

---

3. [It is neither necessary that these hypotheses be true nor even that they be likely, but one thing is sufficient, namely, that the calculation to which they lead agrees with observations.]

credit for them at the moment when their consequences would be receiving the experimental confirmation we have anticipated.

We undertook to write the history of the great laws of statics and dynamics in order for Energetics to be in the position to understand and exhibit the evolution experienced by each of its fundamental principles.

It was known that important reflections on statics were sketched in the manuscript notes of Leonardo da Vinci. Our reading of Leonardo da Vinci and Cardano drew our attention to the unexplored statics of the Middle Ages; and soon, the act of laying bare all the manuscripts on statics at the public libraries of Paris yielded unexpected discoveries in abundance.[4] The Christian Middle Ages had known the writings on statics composed by the Greeks; some of these writings came to it directly and others through the intermediary of Arabic commentaries. But the Latins who read those works were not at all the slavish commentators, devoid of any invention, that people were pleased to depict to us. The remains of Greek thought that they received from Byzantium or from Islamic science did not remain in their minds as in a sterile depository; these relics were sufficient to awaken their attention, to fertilize their intellect. And from the thirteenth century on, perhaps even before that time, the school of Jordanus opened to students of mechanics some paths that antiquity had not known.

At first, the intuitions of Jordanus de Nemore were extremely vague and extremely uncertain; some grave errors were intermixed with some great truths. But soon, the disciples of the great inventor refined the master's thought. The errors were eclipsed and began to disappear; the truths became more precise and firmer, and several of the most important laws of statics were finally established with complete certainty.

Specifically, we owe to the school of Jordanus a principle whose importance was demonstrated, with ever-growing clarity, during the development of statics. Without analogy to the postulates specific to the lever, of which Archimedes's deductions made use, this principle has only a distant affinity to the inexact axiom invoked by Aristotle's *Mechanical Questions*. It affirms that the same motive force can lift different weights to different heights, as long as the heights are inversely proportional to the weights. Applied by Jordanus only to the straight lever, this principle allowed one of his disciples to ascertain the law of the equilibrium of weights on an inclined plane and, by an admirable geometric device, the law of the equilibrium of the bent lever.

---

4. *Les Origines de la statique*, vol. 1 (Paris, 1905–1906).

Descartes took up almost without change what this anonymous mathematician of the thirteenth century had written; and henceforth, from Descartes to Wallis, from Wallis to Bernoulli, and from the former to Lagrange, then to Gibbs, the principle of virtual displacements continued to be extended.

Around the year 1360, Albert of Saxony, a master of arts of the University of Paris, wrote:

> It is not true that every part of a weight tends toward its center becoming the center of the world—which would be impossible. It is the whole that descends in such a way that its center becomes the center of the world, and all the parts tend toward the goal that the center of the whole becomes the center of the world; therefore, they do not impede one another.

This center, this point which, in every weight, tends to place itself at the center of the world is, as Albert repeated on several occasions, the center of gravity.

Therefore, every weight moves as if its center of gravity sought the center of the world—a false idea that, during the seventeenth century, engendered many errors, engaged the greatest geometers, and yielded only after a fierce discussion; but in the meanwhile, it was a fertile idea that imparted new truths to statics. In fact, it immediately gave statics the following proposition: A system of weights is in equilibrium when the center of gravity is as low as possible. Torricelli and Pascal one day accepted that proposition as the foundation of all statics, and it gave rise to the theorem of Lagrange and Lejeune-Dirichlet on the stability of equilibrium.

Leonardo da Vinci, that indefatigable reader, leafed through and meditated endlessly on the writings of the school of Jordanus on the one hand, and the scholastic questions of Albert of Saxony on the other. The former, by acquainting him with the law of the equilibrium of the bent lever, led him to the following memorable law, which governs the composition of concurrent forces: With respect to a point taken on one of the composing forces or on the resulting force, the two other forces have equal moments.[5] Moreover, Albert of Saxony's ideas on the role of the center of gravity allowed him to discover the rule of the polygon of

5. "Léonard de Vinci et la composition des forces concourantes," *Bibliotheca mathematica* 4 (1904); *Les Origines de la statique*, vol. 2; "La Scientia de ponderibus et Léonard de Vinci," *Etudes sur Léonard de Vinci, ceux qu'il a lus et ceux qui l'ont lu*, ser. 1 (Paris, 1906).

support,[6] which Villalpand plagiarized.[7] Thus, we find the origins of several principles essential to statics in the writings composed during the thirteenth and fourteenth centuries.

Was it the same for dynamics?

The dynamics begun by Galileo—and by those who emulated him and his disciples, such as Baliani, Torricelli, Descartes, Beeckmann, and Gassendi—is not an innovation; the modern intellect did not produce it, suddenly and completely, as soon as the reading of Archimedes revealed the art of applying geometry to natural effects.

Galileo and his contemporaries made use of the mathematical skill, acquired in antiquity by the geometers while they practiced their trade, in order to render more precise and to develop a science of mechanics, a science whose principles and most essential propositions had been posited by the Christian Middle Ages. The physicists who taught this mechanics during the fourteenth century at the University of Paris had conceived it by taking observation as their guide; they substituted it for Aristotle's dynamics, convinced of its inability to "save the phenomena." At the time of the Renaissance, the superstitious archaism, which delighted equally in the wit of the humanists and the Averroist habit of retrograde scholasticism, rejected this doctrine of the "Moderns." The reaction against the dynamics of the "Parisians" and for the inadmissible dynamics of the Stagirite was powerful, particularly in Italy.[8] But, in spite of this hard-headed resistance, the Parisian tradition found some masters and savants to maintain it and develop it outside the schools, as well as in the universities. Galileo and his followers were the heirs of this Parisian tradition. When we see the science of Galileo triumph over the stubborn peripateticism of Cremonini, we believe, since we are ill-informed about the history of human thought, that we are witness to the victory of modern, young science over medieval philosophy, so stubborn in its mechanical repetition. In truth, we are contemplating the well-paved triumph of the science born at Paris during the fourteenth century over the doctrines of Aristotle and Averroës, restored into repute by the Italian Renaissance.

6. *Les Origines de la statique*, vol. 2; "La Scientia de ponderibus et Léonard de Vinci," in *Etudes sur Léonard de Vinci*, ser. 1.

7. "Léonard de Vinci et Villalpand," in *Etudes sur Léonard de Vinci*, ser. 1; *Les Origines de la statique*, vol. 2.

8. "La tradition de Buridan et la science italienne au xvi siècle," in *Etudes sur Léonard de Vinci: Léonard de Vinci et les précurseurs parisiens de Galilée*, ser. 3 (Paris, 1913).

No motion can last unless it is maintained by the continuous action of a motive power directly and immediately applied to the mobile. That is the axiom on which all of Aristotle's dynamics rests.

In conformity with this principle, the Stagirite wanted to apply a motive power for transporting the arrow, which continues to fly after having left the bow. He believed he had found this power in the perturbation of air; it is air, struck by a hand or a ballistic machine, which supports and carries forth the projectile.

This hypothesis, which seems to push verisimilitude to the brink of ridicule, appears to have been accepted almost unanimously by the physicists of antiquity.[9] Only one of them spoke clearly against it, and he, living during the final years of Greek philosophy, is almost separated from that philosophy by his Christian faith; we are referring to John of Alexandria, surnamed Philoponus. After having demonstrated what was inadmissible about the peripatetic doctrine of projectile motion, John Philoponus declared that the arrow continues to move without any motor applied to it because the string of the bow has given it an *energy* that plays the role of motive virtue.

The last Greek thinkers and Arabic philosophers did not even mention the doctrine of John the Christian, for whom Simplicius and Averroës had only sarcastic comments. The Christian Middle Ages, in the grip of a naive admiration for the newly discovered peripatetic science, at first shared the Greek and Arabic commentators' disdain for Philoponus's hypothesis; Saint Thomas Aquinas mentions the hypothesis only to warn off those who might be seduced by it.

But, following the condemnations brought forth in 1277 by Etienne Tempier, the Bishop of Paris, against a set of theses upheld by "Aristotle and his followers," there appeared a large movement that liberated Christian thought from the shackles of peripatetic and neo-Platonic philosophy and produced what the Renaissance archaically called the science of the "Moderns."

William of Ockham attacked Aristotle's theory of projectile motion with his customary zeal.[10] He was, however, content in destroying without building; but his critiques restored into repute the doctrine of John Philoponus for some of Duns Scotus's disciples. The *energy*, the motive virtue of which Philoponus spoke, reappeared under the name *impetus*. This hypothesis of impetus—what was impressed into the projectile by

9. "Nicolas de Cues et Léonard de Vinci," in *Etudes sur Léonard de Vinci, ceux qu'il a lus et ceux qui l'ont lu*, ser. 2 (Paris, 1909).

10. "Nicolas de Cues et Léonard de Vinci," in *Etudes sur Léonard de Vinci*, ser. 2.

the hand or the machine that launches it—was taken over by a secular master of the Faculty of Arts of Paris, a physicist of great genius.[11] Toward the middle of the fourteenth century, Jean Buridan took impetus as the foundation of a dynamics that "accords with all the phenomena."

The role that impetus played in Buridan's dynamics is exactly the one that Galileo attributed to *impeto* or *momento*, Descartes to *quantity of motion*, and Leibniz finally to *vis viva*. So exact is this correspondence that, in order to exhibit Galileo's dynamics, Torricelli, in his *Lezioni accademiche*, often took up Buridan's reasons and almost his exact words.

Buridan took this impetus, which remains without change within the projectile unless constantly destroyed by the resistance of the medium and by the action of weight contrary to the motion, to be proportional to the *quantity of primary matter* within the body; he conceived and described that quantity in terms almost identical to those Newton used to define mass. With equal masses, the impetus increases as the speed increases; Buridan prudently abstained from further specifying the relation between the magnitude of the impetus and that of the speed. More daring, Galileo and Descartes affirmed that this relation is reduced to proportionality; thus, they obtained an erroneous estimation for *impeto* and for *quantity of motion*, which Leibniz needed to rectify.

Gravity increases indefinitely, as does the resistance of the medium, and it ends up annihilating the impetus of a mobile thrown upward, since such a motion is contrary to the natural tendency of that gravity. But with a falling mobile, motion conforms to the tendency of gravity. Thus, the impetus must be augmented indefinitely and speed must increase constantly during the motion. Such is, according to Buridan, the explanation for the acceleration observed in the fall of a weight, an acceleration that Aristotle's science already understood but for which the Greek, Arabic, or Christian commentators of the Stagirite had given unacceptable reasons.

This dynamics expounded by Buridan presents in a purely qualitative but always exact fashion the truths that the notions of *vis viva* and work allow us to formulate in quantitative language.

The philosopher of Béthune was not alone in professing this dynamics; his most brilliant disciples, Albert of Saxony and Nicole Oresme, adopted and taught it. The French writings of Oresme allowed it to be understood even by those who were not clerics.[12]

11. "Jean I Buridan (de Béthune) et Léonard de Vinci," in *Etudes sur Léonard de Vinci*, ser. 3.

12. "Dominique Soto et la scholastique parisienne," in *Etudes sur Léonard de Vinci*, ser. 3.

When no resistant medium, when no natural tendency analogous to gravity is opposed to motion, the impetus maintains a constant intensity. The mobile, to which a motion of translation or rotation has been communicated, continues to move indefinitely in the same manner with a constant speed. That is the form under which the law of inertia presented itself to the mind of Buridan; it is the form under which it was received by Galileo.

From this law of inertia, Buridan derived a corollary whose novelty we should admire.[13] The celestial orbs move eternally with a constant speed because, according to the axiom of Aristotle's dynamics, each one is subject to an eternal motor of immutable power. The Stagirite's philosophy required that such a motor be an intelligence separated from matter. The study of the motive intelligences of the celestial orbs was not only the crowning glory of peripatetic metaphysics, it was the doctrine about which revolved all the neo-Platonic metaphysics of the Greeks and Arabs; the scholastics of the thirteenth century did not hesitate to receive this heritage of the pagan theologies into their Christian systems.

Now, Buridan had the boldness to write these lines:

> Since the creation of the world, God has moved the heavens by movements identical to those by which they are actually moved. Hence, he has impressed upon them some impetus by which they continue to be moved uniformly. In effect, these impetuses, encountering no contrary resistance, are never destroyed or weakened. . . . According to this imagination, it is not necessary to posit the existence of intelligences moving the celestial bodies in an appropriate manner.[14]

Buridan expressed this thought in various places; Albert of Saxony formulated it also[15]; and Nicole Oresme, in order to formulate it, made use of this comparison: "Violence excepted, the situation is similar to a man making a clock and letting it go and move by itself."

If we wanted to draw a precise line separating the period of ancient science from the period of modern science, we would have to draw it at the instant when Jean Buridan conceived this theory, at the instant when the stars stopped being perceived as moved by divine beings, when celestial

13. "Nicolas de Cues et Léonard de Vinci," in *Etudes sur Léonard de Vinci*, ser. 2 (Paris, 1909).

14. "Nicolas de Cues et Léonard de Vinci," in *Etudes sur Léonard de Vinci*, ser. 2 (Paris, 1909).

15. "Nicolas de Cues et Léonard de Vinci," in *Etudes sur Léonard de Vinci*, ser. 2.

motions and sublunar motions were admitted as being dependent on a single mechanics.

This mechanics, both celestial and terrestrial, to which Newton gave the form we admire today had been attempting to constitute itself since the fourteenth century. The writings of Francis of Mayronnes[16] and Albert of Saxony[17] during the whole of that century teach us that there were physicists who maintained that one could construct a more satisfactory astronomical system than the one in which the earth is deprived of motion, by assuming the earth mobile and heaven and the fixed stars immobile. Of these physicists, Nicole Oresme developed the reasons for this doctrine[18] with a fullness, clarity, and precision that Copernicus was far from achieving. He attributed to the earth a natural impetus similar to the one Buridan attributed to the celestial orbs. In order to account for the vertical fall of weights, he allowed that one must compose this impetus by which the mobile rotates around the earth with the impetus engendered by weight. The principle he distinctly formulated was only obscurely indicated by Copernicus and merely repeated by Giordano Bruno.[19] Galileo used geometry to derive the consequences of that principle but without correcting the incorrect form of the law of inertia implied in it.

While dynamics was being established, the laws of falling weights were being discovered a few at a time.

In 1368, Albert of Saxony proposed these two hypotheses: The speed of the fall is proportional to the time elapsed from the start, and the speed of the fall is proportional to the path traveled.[20] He did not choose between these two laws. The theologian, Peter Tataret, who taught in Paris toward the end of the fifteenth century, reproduced literally what Albert of Saxony had said. The great reader of Albert of Saxony, Leonardo da Vinci, after having accepted the second of these two hypotheses, rallied to the

16. "François de Meyronnes O.F.M. et la question de la rotation de a terre," *Archivium Franciscannum historicum* 6 (1913).

17. "Un Précurseur français de Copernic: Nicole Oresme (1377)," *Revue générale des sciences pures et appliquées* (1909).

18. "Un Précurseur français de Copernic: Nicole Oresme (1377)," *Revue générale des Sciences pures et appliquées* (1909).

19. "La tradition de Buridan et la Science italienne au xvi siècle," in *Etudes sur Léonard de Vinci*, ser. 3.

20. "Sur la découverte de la loi des chute des graves," *Comptes rendus des séances de l'Académie des sciences* 146 (1908); "Dominique Soto et la scholastique parisienne," in *Etudes sur Léonard de Vinci*, ser. 3.

first. But he was not able to discover the law of spaces traversed by a falling weight; by a reasoning that Baliani took up, he concluded that the spaces traversed in laps of equal and successive times are like the series of whole numbers, whereas, in truth, they are like the series of odd numbers.

The rule that allowed the evaluation of the space traversed in a certain time by a mobile moving in a uniformly varied motion had been known for a long time, however. Whether this rule was discovered in Paris during the time of Jean Buridan or in Oxford during the time of Swineshead, it was formulated clearly in the work in which Nicole Oresme posited the essential principles of analytic geometry.[21] Moreover, the demonstration that serves to justify it is identical to the one Galileo gave for it.

This rule was not forgotten from the time of Nicole Oresme to the time of Leonardo da Vinci; formulated in most of the treatises produced by the thorny dialectics of Oxford, it was discussed in the various commentaries of which these treatises were the object during the fifteenth century in Italy, and then in the various works of physics written at the start of the sixteenth century by Parisian scholasticism.

None of the treatises of which we have just spoken, however, contains the thought of applying this rule to the fall of weights. We encounter that thought for the first time in the *Questions on Aristotle's Physics,* published in 1545 by Domingo de Soto.[22] A student of the Parisian scholastics, most of whose physical theories he received and adopted, the Spanish Dominican de Soto admitted that the fall of a weight is uniformly accelerated, and that the vertical rise of a projectile is uniformly retarded; also, in order to calculate the path traversed in each of these two movements, he correctly used the rule formulated by Oresme. That is, he knew the law of falling weights, whose discovery is attributed to Galileo. Moreover, he did not claim the discovery of these laws; rather, he seemed to be giving them as commonly received truths. No doubt they were accepted at the time by the Paris masters whose lessons de Soto followed. Thus, from William of Ockham to Domingo de Soto, we see that the physicists of the Parisian school posited the foundations of the mechanics that Galileo, his contemporaries, and his disciples developed.

Among those who, before Galileo, received the tradition of Parisian scholasticism, there was none who deserved more attention than Leonardo

21. "Dominique Soto et la scholastique parisienne" and "La Dialectique d'Oxford et la scholastique italienne," in *Etudes sur Léonard de Vinci,* ser. 3.

22. "Dominique Soto et la scholastique parisienne," in *Etudes sur Léonard de Vinci,* ser. 3.

da Vinci. During the time in which he lived, Italy firmly resisted the penetration of the mechanics of the "*moderni*," of the "*juniores*." Among the university masters, even those who leaned in the direction of the terminalist doctrines of Paris merely reproduced, in an abridged and often hesitant form, the essential assertions of that mechanics; they were far from being capable of having it produce any of the fruits of which it was the flower.

Leonardo da Vinci, on the contrary, was not satisfied in admitting the general principles of the dynamics of impetus. He meditated endlessly on these principles and turned them every which way, pressing them in some fashion to deliver the consequences they enclosed.[23] The essential hypothesis of that dynamics was similar to the first form of the law of *vis viva*; da Vinci perceived in it the idea of the conservation of energy, and he found some terms of almost prophetic clarity with which to express that idea.[24] Albert of Saxony had left his reader in suspense between the two laws of falling weights, the one correct and the other inadmissible. After some tentative steps that Galileo also went through, da Vinci came upon the choice of the correct law. He extended it happily to the fall of a weight along an inclined plane.[25] Through a study of composite *impeto*, he attempted the first explanation of the curvilinear trajectory of projectiles, an explanation that was completed by Galileo and Torricelli.[26] He glimpsed the correction that needed to be brought to the law of inertia announced by Buridan, and he prepared for the work that Benedetti and Descartes accomplished.[27]

No doubt, da Vinci did not always recognize the richness of the treasures accumulated by Parisian scholasticism. He set aside some, which would have been complementary to his doctrine of mechanics. He misunderstood the role that impetus must play in the explanation of the accelerated fall of weights.[28] He was unaware of the rule which allows the calculation of the path traversed by a body moving of uniformly accelerated

23. "La tradition de Buridan et la science italienne au xvi siècle," in *Etudes sur Léonard de Vinci*, ser. 3.

24. "Nicolas de Cues et Léonard de Vinci," in *Etudes sur Léonard de Vinci*, ser 2.

25. "La Dialectique d'Oxford et la scholastique italienne," in *Etudes sur Léonard de Vinci*, ser. 3.

26. "Nicolas de Cues et Léonard de Vinci," in *Etudes sur Léonard de Vinci*, ser. 2.

27. "La tradition de Buridan et la science italienne au xvi siècle," in *Etudes sur Léonard de Vinci*, ser. 3.

28. "La tradition de Buridan et la science italienne au xvi siècle," in *Etudes sur Léonard de Vinci*, ser. 3.

motion. It is no less true that the whole of his physics placed him among those whom the Italians of his time called the Parisians.

Moreover, this title was properly given to him. In fact, his principles of physics were derived from an assiduous reading of Albert of Saxony and probably also from a meditation on the writings of Nicholas of Cusa[29]; and Nicholas of Cusa was also an initiate of the Parisian mechanics. Da Vinci is therefore given his proper place among the Parisian precursors of Galileo.

We have just retraced, in broad strokes, the essential laws of equilibrium and motion at their infancy. On occasion, we have described some portions of physics at the time when that science had reached adolescence. Thus, we have inquired into the sources of the hydrostatic theories of Pascal,[30] detailed the role that Mersenne played in the discovery of the weight of air,[31] and sketched the genesis of the doctrine of universal attraction.[32] Now, we did not see any essential principles proceed from the desire to resolve the bodies we perceive and touch into imperceptible but simpler bodies; we saw none that had as its aim to explain sensible motions by means of hidden motions. Atomism did not contribute to their formation in any way. All of them were born from the desire to formulate some general rules whose consequences "saved the phenomena." Thus, the history of the development of physics has come to confirm what the logical analysis of the methods used by that science had taught us. From the former and from the latter, we have gained a renewal of faith in the future fruitfulness of the method of Energetics.

---

29. "Nicolas de Cues et Léonard de Vinci," in *Etudes sur Léonard de Vinci*, ser. 2.

30. "Le Principe de Pascal; essai historique," *Revue générale des sciences pures et appliquées* (1905).

31. "Le P. Marin Mersenne et la Pesanteur de l'air—Première partie: Le P. Mersenne et le poids spécifique de l'air," and "Le P. Marin Mersenne et la Pesanteur de l'air—Seconde partie: Le P. Mersenne et l'expérience du Puy-de-Dôme," *Revue générale des sciences pures et appliquées* (1906).

32. *La Théorie physique, son objet et sa structure.*

12

Some Reflections on German Science[1]

In February and March of 1915, in the midst of the French conflict with Germany during World War I, Duhem gave a series of lectures on German science and the German mind, under the auspices of the Catholic Student Association of the University of Bordeaux. Duhem then published the lectures as a book, La Science allemande, *including this article, published earlier that year in* La Revue des deux mondes. *The dedication of the work gives us its purpose: "To the Catholic students of the University of Bordeaux: I dedicate these lectures, written at their request and given under their auspices. With God's help, may these humble pages preserve and promote the clear intellect of our France in them and in all their schoolmates!" (p. iii).*

The first few pages are even more revealing. Duhem refers to the Occupation. He talks about a friend telling him that it was not just the territory of France that was invaded; foreign thought has reduced French thought to a position of servitude. Duhem is asked to sound the charge that will liberate the soul of the motherland, and he takes on the obligation willingly: "My combat station has been assigned to me. I accept it. The station is without danger; thus, it will be without glory. I cannot spill my blood, but I will spill everything my heart contains of devotion. I come before you to play my humble part for the national defense" (p. 4).

La Science allemande *is evidently an example of wartime propaganda, Duhem's personal contribution to the war effort. But it is not a one-dimensional example. The book sold very quickly, and there were even inquiries about the*

1. ["Quelques réflexions sur la science allemande," *Revue des deux mondes* 25 (1915): 657–686. Also in *La Science allemande* (Paris: Hermann, 1915).]

possibility of translating it into German. The Germans were not offended by it, possibly because, in the climate of the times, Duhem's book was a rarity; the book was actually complimentary about German science, praising such scientists as Karl Friedrich Gauss and Hermann Helmholtz. The lasting significance of the work lies, however, in its elaboration of the themes of French and English minds, from "English School and Physical Theories" and The Aim and Structure of Physical Theory, *with the addition of the German mind.*

I

Formerly, we have tried to describe the stamp which impresses so particular and salient a character on English theoretical physics. Today we want to try, in a similar manner, to uncover the marks proper to the doctrines of mathematics or physics made in Germany.

Such an attempt must guard itself against claiming to come to any rigorous conclusion. Taken in its essence, considered under its perfect form, science must be absolutely impersonal. Since no discovery in it would bear the signature of its author, neither would anything allow one to say in what country the discovery saw the light of day.

But this perfect form of science could not be obtained except by a precise separation of the various methods concurring in the discovery of truth. Each of the many faculties that human reason puts into play when it wishes to know more and better would have to play its role, without anything being omitted, without any faculty being overlooked.

This perfect equilibrium between the many organs of reason does not occur in any single person. In each of us, one faculty is stronger and another weaker. In the conquest of truth, the weaker will not contribute as much as it should and the stronger will take on more than its share. The science produced by this poorly apportioned work will not show the harmonious proportions of its ideal exemplar. The faulty development of some parts corresponds with the excessive growth of some others. It is in these deformities alone that we can recognize the turn of mind of the author.

It is these too that will frequently allow us to name the set of people who produced a particular theory.

The body of each person deviates from the ideal type of the human body by the exaggerated proportions of one organ, by the diminished proportions of another. These kinds of small monstrosities that distinguish us from one another are also those that physically characterize the various nations: this or that exaggerated or arrested development is particularly frequent among this or that people.

What is said of the body can be repeated of the mind. To say that a people has its own particular mind is to say that, quite frequently, in the reason of those who make up this people, one faculty is developed more than is suitable and another faculty does not have its full breadth and all its force.

Two conclusions follow immediately from that:

First, judgments about the intellectual form of a people will be able to be verified frequently, but they will never be universally true. Not all the English are of the English type. Even more so, theories conceived by the English will not show all the characteristics of English science. There will be some works that could just as well be taken for French and German works. In return, there will be some intellects in France that think in the English mode.

Second, if the national character of certain authors is perceived in the doctrines they have created or developed, it is because this character has molded the way in which these doctrines diverged from their perfect type. It is by its defects—and by its defects alone—that science, diverging from its ideal, becomes the science of this or that people. One can therefore expect that the marks of the genius proper to each nation be particularly prominent in works of the second order, the products of lesser thinkers. Quite often the great masters possess a reason in which all the faculties are so harmoniously proportioned that their perfect doctrines are exempt from all individual character, as from all national character. There is no trace of the English mind in Newton, nothing of the German mind in the work of Gauss or that of Helmholtz. In such works, one no longer sees the genius of this or that people but only the genius of Humanity.

II

"Principles are felt, propositions proved," said Pascal,[2] who must always be cited when we claim to be speaking about scientific method. In every science assuming the form we call rational—or, better yet, the form that might be called mathematical—we must, in fact, distinguish two strategies: the one that captures principles and the one that arrives at conclusions.

2. [B. Pascal, *Pensées*, VI, 110 (282), trans. Krailshimer (London: Penguin, 1966), p. 58.]

The method that arrives at conclusions from principles is the deductive method followed with the most rigorous precision.

The method that leads to the formulation of principles is much more complex and difficult to determine.

Are we dealing with a purely mathematical science? Common experience is the matter from which induction derives axioms. Deduction draws out all the truths contained in these universal propositions. Now, the choice of axioms is an operation of extreme delicacy. Axioms must be sufficient to satisfy all the propositions of science we wish to extract. The chain of reasoning must not suddenly have its continuity broken and its rigor compromised because a principle necessary for its progress has remained hidden in the data of experience and has not yet been formulated explicitly. It is equally necessary that there not be too many principles, that a simple corollary of other axioms not be given as an axiom. If we follow the history of the axioms of geometry from Euclid's *Elements* to Hilbert's works, we will see to what extent the choice of principles for a mathematical science is a minute and complicated task.

More complex yet is the choice of hypotheses on which will rest the whole edifice of a doctrine belonging to experimental science, of a theory of mechanics or physics.

Here the matter that must provide principles is no longer common experience, that which every man spontaneously exercises from the time he leaves infancy; it is scientific experiment. Common experience furnishes autonomous, rigorous, definitive data to the mathematical sciences. The data of scientific experiments are only approximate; the continuous improvement of instruments perfects them and modifies them endlessly, while fortunate discoveries every day come to enlarge the treasury with some new fact. Finally, far from being autonomous or immediately intelligible in themselves, propositions that formulate the result of an experiment in physics or chemistry have meaning only if accepted theories provide their translations.

Physicists must extract their principles from this inextricable web in which the data of sensation lie tangled up, assisted by more and more complicated instruments, by interpretations furnished by changeable theories subject to caution, and sometimes by the very theory they propose to modify. In the inspection of this confused mixture, they must guess at the general propositions from which deduction will derive conclusions conformable to the facts.

They would find in deductive method only an extremely rigid and not very penetrating helper for accomplishing this work. They need a more supple and less rigid method than that. More so than mathematicians, physicists will need a faculty distinct from the geometrical mind in order to choose their axioms. They will have to appeal to the intuitive mind.

III

The intuitive mind and the geometrical mind do not proceed in the same fashion.

The procedure of the geometrical mind obeys inflexible rules imposed on it from outside. Each of the propositions it unfolds one from the other has its place marked in advance by a necessary law. To escape this law, ever so little, and to pass from one judgment to another by jumping over some intermediary required by deductive method, is, for this mind, to lose its power, which consists entirely in rigor. The word *linkage* comes to our lips as soon as we want to define the order in which syllogisms succeed one another. In fact, the chain that ties together such reasoning does not allow for any freedom.

If the geometrical mind owes all the force of its deductions to the rigor of its procedure, the penetration of the intuitive mind comes entirely from the spontaneous suppleness with which it operates. No unchangeable precept determines the path which its free endeavors will follow. Sometimes one sees it cross the abyss that separates two propositions with a bold leap. Sometimes it slips into and insinuates itself among the numerous objections preventing access to a truth. It is not that it proceeds without order, but that it prescribes for itself the order it follows; it constantly modifies it in light of circumstances and events in such a fashion that no precise definition could pin down its meanders and unexpected leaps.

The procedure of the geometrical mind calls forth the idea of an army marching past on review. The various regiments are aligned with impeccable regularity. Each person holds exactly the rank allotted to her by a strict order. Each feels held there by an iron discipline.

The procedure of the intuitive mind recalls rather that of the sharpshooter assaulting a difficult position. At one moment, he leaps suddenly; at another he creeps stealthily by the obstacles erected on the slope. There, also, each soldier follows an order. But no part of this order is explicitly formulated except the goal of winning. The free interpretation given by each of the assailants about how best to implement this goal makes the various motions tend toward the specified aim.

Does not this comparison between the procedure of the intuitive mind and that of the geometrical mind allow us to predict the proper character of German science, what will distinguish it in particular from French science? No doubt, with the greater number of the French who cultivate the intuitive mind, science will be marked by its excessive application; not satisfied with the role given it, impatient with the heavy slowness of the geometrical mind, the intuitive mind will at times encroach upon the prerogatives of the latter. No doubt, we must also expect to see that German science is often deficient in the intuitive mind and concedes to the geometrical mind that which is not its legitimate possession.

Let us glance at some of the works that have established the reputation of German science and see if the predominance of geometrical mind over intuitive mind is not easily recognized there.

IV

The geometrical mind could be still better called the algebraic mind. There is no part of science, in fact, in which the deductive method has a greater role than in this vast generalization of arithmetics called algebra or analysis. The axioms on which it rests consist of a small number of simple propositions about whole numbers and their addition. The intuitive mind did not have to make a great effort to disengage them from the most common experience. The innumerable truths of which the science of algebra is constructed can be derived from these axioms through the most rigorous series of syllogisms conceivable.

The faculty of following without fail the most minute rules of logic in the course of long and complicated reasoning is not, however, the only faculty that comes into play in the construction of algebra. Another faculty takes an essential part in this work. It is the one by which mathematicians, in the presence of a complex algebraic expression, easily perceive the various transformations allowable by the rules of calculation which they can make it undergo and, thus, reach the formulas they wanted to discover. This faculty, analogous to that of a chess player preparing a clever move, is not a power of reasoning but an aptitude for combining things.

Among German mathematicians, there are doubtless those who have possessed this aptitude for combining the operations of algebraic calculation to a high degree. But it is not in this that the analysts from beyond the Rhine have excelled. The grand masters of this art—such as Hermite, Cayley, and Sylvester—are more easily found in France and, above all, in England. It is by its power of deducing with the most extreme rigor, of following without the least lapse the longest and most complicated chains of reasoning, that German algebra has marked its superiority. It is by this power that Weierstrass, Kronecker, and Georg Cantor, for example, have displayed the force of their geometrical minds.

By this absolute submission of their geometrical minds to the rules of deductive logic, German mathematicians have usefully contributed to the perfection of analysis. The algebraists who before them had shone in other nations had too readily and improperly trusted the intuitions of the intuitive mind. Thus, they had often come to formulate as demonstrated some truths that were in fact only conjectured. At times propositions had even been hastily given as correct when they were not. German science has greatly contributed to ridding the field of algebra of all paralogisms.

Let us cite only one example out of a thousand. By means of too prompt and too hasty an intuition, the intuitive mind believed it knew that every continuous function had a derivative. Pressing the geometrical mind improperly, it had persuaded the latter to accept some apparent demonstrations of that proposition. In forming continuous functions which never have derivatives, Weierstrass has shown how dangerous the temporary abandonment of rigor could be in the course of an algebraic deduction.

Thus, the extreme rigor of the geometrical mind has great advantages for the progress of algebra. It also presents serious inconveniences. Overly cautious of avoiding or resolving objections that are only trifles, it encumbers science with otiose and tedious discussions. It suffocates the spirit of invention. In fact, before forging the chain that ought to connect a new truth to principles by means of proven links, we must have first perceived that truth. The intuition that precedes demonstration in all mathematical discovery is an endowment of the intuitive mind. The geometrical mind does not know it, and, in the name of rigor, it readily denies it the right to function. There are even some geometers in Germany, such as Felix Klein, who have come to reassert the place of intuitions proper to the intuitive mind in the domain of algebraic method because they were uneasy about the dangers that overly exclusive use of the geometrical mind causes for the faculty of invention.

V

Algebra subjects reason to the iron discipline of the laws of syllogism and rules of calculation. No science is better adapted to the German mind, proud of its mathematical rigor but deprived of intuition. So Germans tried to give every science a form which, as much as possible, recalled that of algebra. For example, in their hands, geometry was reduced to nothing but a branch of analysis.

By inventing analytic geometry, Descartes had already referred the study of shapes traced in space to discussions of algebraic equations. He taught us to express each point in space by using three numbers, the *coordinates* of the point. For a point to be found on a given surface, it is necessary and sufficient that its three coordinates satisfy a certain equation. All information about the algebraic properties of the equation is, from then on, information about the geometrical properties of the surface; the inverse is also true. Thus, those who are more skilled at combining formulas than at considering assemblages of lines and surfaces will be great geometers by virtue of the fact that they are skilled algebraists.

Even after Descartes's work, however, the reduction of geometry to algebra was not absolute. In order to attribute three coordinates to a point in space, it was still necessary to refer to some geometrical propositions, to the most elementary theorems about straight lines and parallel planes. Simple as these propositions may have been, they entailed the acceptance of all the axioms Euclid required at the beginning of the *Elements.* Now, for those whose geometrical minds suffer from the slightest lack of rigor, this adherence to Euclid's axioms is scandalous.

The axioms required by a science of reasoning should not merely agree among themselves without any shadow of contradiction; they should, furthermore, be as few as possible in number. Hence, they should be independent of one another. In fact, if one of them could be demonstrated by means of the others, it would be deleted from the number of the axioms and placed among the theorems.

Now, are Euclid's axioms truly independent of one another? This is a question that has long worried geometers. Among these axioms is one that is the basis for the theory of parallel lines. Many have thought to recognize it as a simple corollary of the other postulates formulated by Euclid, so we have seen repeated attempts at its demonstration. But a somewhat perspicacious critic has always discovered a vicious circle in each of these attempts.

The question was given another, more ingenious, slant by Gauss, Bolyai, and Lobachevskii. These mathematicians applied themselves to unfolding the series of propositions that could be established by accepting all the axioms formulated by Euclid, except the parallel postulate. They thought, that if one could follow to infinity the series of consequences of those axioms without assuming the truth of the dubious postulate, and without ever encountering a contradiction, it would be because the adoption of these principles does not necessarily require the truth of the parallel postulate. Henri Poincaré has shown the well-foundedness of this thought conceived by Gauss, Bolyai, and Lobachevskii. He has shown that if the non-Euclidean geometry constructed by these mathematicians could ever end up with two mutually contradictory propositions, it would be because Euclidean geometry itself provided two incompatible theorems.

To recognize whether all of Euclid's axioms are truly independent of one another is a question under the jurisdiction of the geometrical mind; and with Gauss, Bolyai, Lobachevskii and their successors, the geometrical mind has fully resolved it. But to decide whether Euclid's postulate is true is a question that the geometrical mind, left to itself, cannot answer. It must be aided by the intuitive mind.

The truth of geometry does not consist merely in the absolute independence of the axioms from one another, or in the impeccable rigor with which theorems are deduced from axioms. It also, and above all, consists in the agreement between the propositions forming this logical chain and the knowledge given to our reason about space and the shapes that can be traced in it by that lengthy experience called common sense. It belongs to the geometrical mind to verify the precision of the deduction by which all the propositions are derived from one another. But it has no means of recognizing if they are or are not conformable with what we know, prior to all geometry, about plane or solid figures. This latter task is the concern of the intuitive mind.

Now, prior to all geometry, one of the first truths we can formulate on the subject of space is that it has three dimensions. When the intuitive mind analyzes this proposition in order to grasp exactly what is intended by it, does it discover that it has this meaning: "To every point in space there correspond three numbers that are its coordinates"? Not at all. What it finds is that, in attributing three dimensions to space, the nonmathematician claims to say this: All bodies have length, breadth, and depth. And when the assertion is pushed, the intuitive mind recognizes that it is equivalent to this other assertion: Every body can be contained exactly in a container of determined size, whose shape is called a rectangular parallelepiped by the geometer. The geometrical mind then comes along to demonstrate that propositions about rectangular parallelepipeds, judged true by the intuitive mind, entail Euclid's celebrated postulate.

By searching the treasury of truths about the sizes and shapes that the most common experience amasses, the intuitive mind comes across these propositions: One can represent a plane figure by drawing a solid figure by sculpture, and the image can perfectly resemble the model even though it is of another size. This is a truth that deer hunters on the banks of the Vézère during Paleolithic times never doubted. Now, that shapes can be similar without being equal presupposes the correctness of Euclid's postulate, as the geometrical mind demonstrates.

To recognize thus the large part that belongs to the intuitive mind in the verification of the axioms of geometry is not to the taste of German science. The latter makes little of the agreement between the propositions of geometry and the knowledge derived from common sense, since this agreement cannot be ascertained by the geometrical mind. It would have the truth of geometry consist exclusively in the rigor of deductive reasoning by which theorems derive from axioms; and, in order not to be exposed to compromising this rigor by borrowing some information from sensible experience, it would reduce geometry to absolutely nothing more than a problem of algebra.

For it, a point will be *by definition* the set of three numbers. Let the values of the three numbers vary continuously in such a set; then it will be said that the point generates a space. The distance between the two points will be *by definition* an algebraic expression consisting of the three numbers of the first set and the three numbers of the second set. No doubt, this algebraic expression is not taken absolutely at random; it is chosen in such a way that some of its algebraic properties are expressed by phrases similar to those setting out certain geometrical properties attributed by common sense to the distance between two points. But these properties will be as few as possible, lest the intuitive mind find in them the pretext for penetrating into the domain of the science to be constructed. Algebraic calculations, called geometry, will then be developed.

Perhaps the intuitive knowledge that reason provides about plane figures and bodies might still find a way of insinuating itself into the stitches of the deductive net woven by this algebra. A new precaution will be taken against this dreaded intuition. Our intuition does not know any space without two or three dimensions; to articulate propositions that might speak of a space of more than three dimensions would be to utter intuitively meaningless words. These are precisely the kinds of propositions that will constantly be formulated. What will be called a point will not be a set of three numbers, as was assumed, but a set of n numbers. The value of the whole number represented by n will not be specified. That value can be greater than three; it can be as great as one wishes. The set of n numbers is, it will be said, a point in a space of n dimensions.

That is how the powerfully geometrical genius of Bernhard Riemann went about writing a chapter of profound algebra which he entitled *On the Hypotheses that Serve as the Foundations of Geometry* (*Über die Hypothesen welche der Geometrie zu Gründe liegen*).

We have indicated with what minute care the intuitive knowledge of lines and surfaces had been set aside in the composition of that doctrine. Is it not astonishing that the corollaries this algebra achieves, which it expresses with words borrowed from geometry, run counter to the propositions that intuitive knowledge of space regards as the most certain? When it affirms, for example, the meeting at a finite distance of any two lines whatsoever in the same plane, does it not deny the very existence of parallels?

Riemann's doctrine is a *rigorous algebra* because all the theorems it formulates are exactly deduced from postulates it sets out. Therefore, it satisfies the geometrical mind. It is not a *true geometry* since, in positing its postulates, it does not bother to make the corollaries agree in all points with the judgments, derived from experience, that make up our intuitive knowledge of space. Thus, it offends common sense.

VI

Riemann's treatise on the foundations of geometry is one of the most justly celebrated works in German science. It seems to be a remarkable example of the procedure by which the German geometrical mind transforms every doctrine into a kind of algebra.

This mind assigns extremely unequal shares to the two methods by whose help every science of reasoning progresses. It develops, with as much breadth as detail, the deduction by which the corollaries are derived from principles. It suppresses or reduces to the smallest extent the set of inductions and conjectures by which the intuitive mind was able to disengage principles from the data of experience.

The hypotheses on which any theory of mechanics or mathematical physics rests are fruits whose ripeness has been prepared for a long time. The data of common observation, results of scientific experiment assisted by instruments, ancient theories now forgotten or rejected, metaphysical systems, and even religious beliefs have contributed to them. Their effects have intersected and their influences have mixed in so complex a manner that a great intuition of mind, sustained by a deep knowledge of history, would be required to make out the essential direction of the path that has led human reason to the clear perception of a principle of physics.

Now, let us examine some of the lessons of such a scientific algebra, in which Gustav Kirchhoff has set forth the various doctrines of mathematical physics. We find no trace of that long and complicated elaboration that has preceded the adoption of principles. Each hypothesis is presented *ex abrupto*, under the very abstract and general aspect which it has taken on after many evolutions and transformations, without a word allowing us to suspect the indispensable preparation. A Frenchman who had heard Kirchhoff in Berlin recently repeated to me the formula by which the German professor used to present each new principle: "We can and will posit . . . (*Wir können und wollen setzen* . . .)" Provided that no contradiction forbids our assumption for the pure logician, we prescribe it as a decree of our free will. This act of will, this choice of our pleasure, is substituted, so to speak, purely and simply for all the work that, over the course of the ages, the intuitive mind had to complete. It leaves nothing subsisting in science except that which submits to the rigid discipline of the geometrical mind. A theory of physics, beginning with postulates freely formulated, is no more than a series of algebraic deductions.

Kirchhoff is not alone in treating mechanics and physics in this way. Those who have followed his lectures imitate his method. For example, can one imagine a more absolute algebraism than the one that inspires Heinrich Hertz when he claims to construct mechanics? The disposition, at a given instant, of the various bodies making up the system under study is known when the values taken by a certain number n of magnitudes is known. For fear that experimental intuition might suggest to us some property of this mechanical system, we quickly lose sight of and forget the bodies that form it, dodge the intuition, and consider only a point whose coordinates, in the space of n dimensions, will be precisely these n values. We agree that this point, which is itself only an algebraic expression, a word of geometric consonance taken to designate a set of n numbers, changes from one instant to another, in such a way as to render to a minimum a certain magnitude represented by an algebraic formula. From this convention, so perfectly algebraic in nature, so fully arbitrary in appearance, we deduce with extreme rigor the consequences that the calculation can derive from it and say we are generating mechanics.

Doubtless, the postulate formulated by Hertz is not as arbitrary as it appears. It was arranged in such a way that its algebraic statement summarized and condensed everything that intuitions, experiments, and discussions had disclosed to students of mechanics concerning the law of inertia and the connections by which bodies inhibit one another in their motions, from Jean Buridan to Galileo and Descartes, and from the latter to Lagrange and Gauss. But Heinrich Hertz has not preserved the least reminder of that previous elaboration in the absolutely precise and rigorous exposition he gives us of mechanics. He completely and systematically abstracts it away, so that the fundamental principle of science takes the imperious form of a decree brought forth by a freely authoritarian alge-braist: *sic volo, sic jubeo, sit pro ratione voluntas* [as I will, so I order; let my will substitute for reason].

Moreover, such a manner of proceeding can produce very good results in certain cases.

By patiently unraveling the complex knot of operations that have slowly produced a hypothesis of physics, the intuitive mind sometimes deludes itself about the role it has played. It comes to imagine that it has fashioned a work of the geometrical mind. It wrongly takes for a categorical demon-stration of a proposition the series of considerations, with transitions deli-cately managed, through which it has little by little prepared the mind to receive that proposition. Our French physics has too often and too long given itself to that pipe dream. It is important to put reason on guard against this misapprehension, not to allow it to believe that a principle of physics is demonstrated only by making it seductive. It is good to remind it that, from the point of view of deductive logic, the hypotheses of physics appear as propositions imposed by no reasoning. Scholars formulate them as they please, led only by the hope of deriving from them corollaries in conformity with the data of experience, and they propose them for our acceptance because the condensation of a multitude of experimental laws and a small number of theoretical postulates appears to them, in the words of Ernst Mach, a fortunate economy of thought. The pure algebraism of German theories is marvelously suited for this task.

But what does that say? Simply that an exposition of physics in which the intuitive mind has exaggerated its power is corrected by another expo-sition in which the intuitive mind has been driven out with too much force—in other words, that an excess often finds its remedy in a contrary excess. Belladonna and digitalis neutralize each other's effects. Neverthe-less, they are both poisonous plants.

VII

We risk giving in to a serious eccentricity if we posit the hypotheses of a theory of mechanics or physics without any concern for the considerations by which the intuitive mind could prepare for our acceptance of them. We are liable to produce doctrines that contradict the universally received teachings of common sense.

German science cares little for the requirements of common sense. It is not displeased to oppose them directly. Bernhard Riemann's geometric theory has already let us recognize this. At the base of the systems it constructs with an apparatus so scrupulously designed, German thought appears at times to take malicious pleasure in positing some assertions which, for the intuitive mind, may be the occasion for scandal, if these assertions should contradict the most assured principles of logic. What a delicious exercise for a geometrical mind that despises the intuitive mind and good sense to place a formally contradictory proposition among the axioms and to derive from such a principle a whole set of corollaries by a series of absolutely conclusive syllogisms!

From early on, there were men in Germany who persevered at this task.

Nicholas of Cusa, the first original thinker to be counted as German, wrote his treatise, *De docta ignorantia* [*On Learned Ignorance*], before the middle of the fifteenth century. In order to provide a basis for the philosophical edifice he was going to erect, the "German Cardinal" made this assertion, whose contradictory character leaps to one's eyes: In every order of things, the maximum is identical to the minimum. Then, on this basis, deductive method allowed him to construct a whole metaphysics.

The nineteenth century produced in Germany an attempt no less strange than that of Nicholas of Cusa. Hegel rested his whole philosophical system on the assertion of the identity of contraries; and the great success that Hegelianism has known in universities beyond the Rhine marks the extent to which the geometrical mind of the Germans, far from being shocked by this defiance of common sense, took pleasure in that great feat of purely deductive method.

A being whose nature consists in feeling dominated by an iron discipline finds its happiness in obeying orders without discussion. The more strange, even revolting, the order, the more joyful the obedience. This explains the cheerful submission with which the geometrical mind of someone such as Nicholas of Cusa or Hegel unfolds the consequences of an absurd principle. Moreover, German metaphysicians have not been alone in supplying examples of this intellectual submission that disconcerts us. We have seen mathematicians unravel entire geometries in which one of the least discussible Euclidean axioms was replaced by its contradictory; and the authors of these deductions seemed to take pleasure in it, in proportion to the inconceivability and ridiculousness of the conclusions as judged by good old common sense.

These mathematicians, however, use a geometry in conformity with this good old common sense every time they happen to measure a body or draw a figure in everyday life.

A similar inconsistency is not at all rare when the geometrical mind claims to do without the help of the intuitive mind. Isolated from common sense, the geometrical mind can truly reason and deduce without end but is incapable of directing action and maintaining life. Common sense rules as master in the domain of facts. Between this common sense and discursive science, the intuitive mind establishes a perpetual circulation of truths that extracts from common sense the principles from which science will deduce its conclusions, and it takes up among these conclusions all that can increase and perfect common sense.

German science does not know this continual exchange. Submissive to the rigorous discipline of the purely deductive method, theory follows its regular march without any care for common sense. Common sense, on the other hand, continues to direct action without theory appearing in any way to sharpen its primitive and crude form.

Does not the idealist philosopher provide raw evidence for this absence of all interpenetration between science and life? In his chair at the university, he denies all reality to the external world because his geometrical mind has not encountered this reality at the end of a conclusive syllogism. An hour later, at the tavern, he finds a fully assured satisfaction in the weighty realities of his sauerkraut, his beer, and his pipe.

With Germans, pure geometers deprived of intuitive mind, life does not guide science and science does not illuminate life. Hence, in his magnificent study *L'Allemagne et la guerre* [*Germany and the War*], Emile Boutroux could write:

> Their science, an affair of specialists and scholars, has not been able to penetrate their soul and influence their character. . . . Apart from notable exceptions, of course, consider in the tavern, in his dealings with ordinary life, in his amusements, the learned professor who excels at discovering and gathering together all the materials for a study and bringing forth from it, through mechanical operations and without the least appeal to judgment and good common sense, solutions completely based on texts and on reasonings. What disproportion there often is between his science and his degree of education! What vulgarity of taste, sentiment, and language, what rough behavior is displayed by this man, whose authority is inviolable in his specialty! . . . With Germans, the scholar and the man are only too often strangers to one another.[3]

The same thing holds for German science that holds for German scientists. The absence of intuitive mind leaves a gaping abyss between the development of ideas and the observation of facts. Ideas are deduced from

3. Emile Boutroux, *L'Allemagne et la guerre* (Paris, 1914) [first published in *Revue des deux mondes* 23 (15 October 1914): 387–388].

one another, proudly contradicting common sense, from which they borrow nothing. Common sense manipulates realities and notices facts by its own means without concern for a theory that ignores it or goes against it. Such is the spectacle which, quite often, the physics from beyond the Rhine presents to us.

VIII

German theories of electrical phenomena provide examples of this incoherent duality.

There is a particularly difficult and complicated doctrine in mathematical physics: the theory of electricity and magnetism. The genius of people such as Poisson and Ampère had set out the principles of this doctrine with wholly French clarity. Before the middle of the nineteenth century, the work of these great men had served as a guide for the work that the most illustrious German physicists, such as Gauss, Wilhelm Weber, and Franz Neumann, had accomplished in order to complete it. All these efforts, inspired by the intuitive mind and disciplined at the same time by the geometrical mind, had built up one of the most powerful and most harmonious theories of physics ever admired. For several years now, this doctrine has been completely reversed by the exclusively geometrical mind of the Germans.

The origin of this reversal does not reside in Germany; it must be sought in Scotland.

The Scottish physicist James Clerk Maxwell seemed to be haunted by two intuitions.

First, insulating bodies, those Faraday called *dielectrics*, must play a role, with respect to electrical phenomena, comparable to conducting bodies. It is reasonable to constitute an electrodynamics for dielectric bodies similar to the one Ampère, W. Weber, and F. Neumann constituted for conducting bodies.

Second, electrical actions must be propagated within a dielectric body in the same way that light is propagated within a transparent body. The velocity of electricity and the velocity of light ought to have the same value for the same substance.

Hence, Maxwell sought to extend the equations of the mathematical theory of electricity to dielectrical bodies and to put these equations into such a form that the identity between the propagation of electricity and the propagation of light could be clearly recognized in them. But the best-established laws of electrostatics and electrodynamics in no way lent themselves to the transformation dreamed up by the Scottish physicist. Throughout his life, one way or another, he tried to reduce these rebellious equations so as to extract from them the propositions he had glimpsed and which he, with his marvelous genius, conjectured to be close to the truth. None of his deductions was viable, however. If he finally obtained the desired equations, it was, with each new attempt, at the price of flagrant paralogisms—indeed, of serious mistakes in calculation.

Maxwell's work was certainly not a German work; to capture the truths revealed by his intuition, Maxwell, the most impulsive and boldly intuitive mind since Fresnel, imposed silence on the most justified complaints of the geometrical mind. The geometrical mind, in turn, had the right and duty to let itself be heard. Maxwell had proceeded to his discoveries by a path broken by precipices impassable to any reasoning respectful of the rules of logic and algebra. It fell to the geometrical mind to trace out an easy route by which one could rise to the same truths without lacking any rigor.

This indispensable work was carried out by a German, but by a German whose genius seemed to be exempt from the defects of the German mind. Hermann von Helmholtz showed how, without abandoning the proven truths long acquired by electrodynamics, without in any way colliding with the rules of logic and algebra, one could nevertheless attain the end the Scottish physicist had proposed. For that purpose, it was sufficient not to impose a velocity rigorously equal to the one Maxwell assigned to the propagation of electrical actions; the velocity imposed was very near the one Maxwell wanted.

The intuitive mind and the geometrical mind were satisfied by Helmholtz's attractive theory. Without denying the electrodynamics constructed by Ampère, Poisson, W. Weber, and F. Neumann, it enriched it with everything that was true and fruitful in Maxwell's views. This theory, so satisfying to all harmoniously constituted reason, was proposed by a German, and that German, whose discoveries in the most varied domains rendered him famous, enjoyed a great and legitimate renown in his own land. The theory found no favor in Germany, however. Even Helmholtz's students made nothing of it. It was one of them, Heinrich Hertz, who gave to Maxwell's thought the form that German science perfected from then on, the geometrical mind having rigorously driven out the intuitive mind.

Objections as numerous as they were serious barred the way to the various methods by which Maxwell tried to justify the equations he hoped to obtain. There was a simple but almost brutal way of clearing away all these objections with one blow. The way was no longer to look upon Maxwell's equations as objects of demonstration, no longer to make of them the terms of a theory for which the commonly received laws of electrodynamics must serve as principles. It was to posit them at the outset as postulates of which algebra had only to unfold the consequences. That was what Hertz did. "Maxwell's theory," he proclaimed, "is no more than Maxwell's equations." The German geometrical mind took singular pride in this manner of acting. In fact, there was no need to have recourse to the intuitive mind to deduce corollaries from equations whose origin is no longer in question; algebraic calculation sufficed.

It goes without saying that this way of proceeding does not satisfy common sense. In fact, Maxwell's equations not only collide with the teachings of a learned and complicated physics, they directly contradict truths accessible to everyone: The mere existence of a permanent magnet is inconceivable for anyone who regards these equations as universally and rigorously true. Hertz explicitly recognized this, and so did Ludwig Boltzmann. Neither, however, saw in this a sufficient reason for denying Maxwell's equations the title of axioms. Now, it is not only in physics laboratories that one finds permanent magnets, lodestones, needles, bars, and horseshoes of magnetized steel. On the bridge of every ship, the binnacle, or ship's-compass stand, contains some magnets; one even comes across them among children's toys. Common sense is assuredly within its rights when it prohibits the geometrical mind from denying their existence.

Permanent magnets are also found among the instruments used by physicists who, on Hertz's advice, treat Maxwell's equations as orders and whose reason obeys these equations without examining their title to such an authority. With the help of instruments provided with permanent magnets, these physicists conduct many experiments. They invoke results of these experiments when, in specific cases, they claim to apply the corollaries of Maxwell's equations. These results then tell them what value it is appropriate to attribute to electrical resistance or to the coefficient of magnetization. But how can they use permanent magnets at the very moment when they invoke a doctrine whose axioms make the existence of such bodies absurd?

Such inconsistency naturally follows a lack of the intuitive mind. Reduced to its own powers, the geometrical mind never knows how to apply its deductions to the data of experience. Between the abstractions theoreticians consider in their reasoning and the concrete bodies observers manipulate in the laboratory, it is the intuitive mind alone that grasps an analogy and establishes a correspondence. The link between theoretical physics and experimental physics is felt, not concluded.

If a theory has been constructed following the laws of a sound method, if the geometrical mind and the intuitive mind have played their legitimate roles, the connection between the equations that the geometrical mind analyzes and the facts that common sense establishes will be convenient and solid. It will result from the very operations by which the intuitive mind has derived the hypotheses that carry the theory, from the teachings of experience. But if the foundations of the theory have not been extracted from the bowels of reality by the intuitive mind, if they are algebraic postulates arbitrarily posited by the geometrical mind, there will no longer be any natural contact between the consequences of theory and the results of experience. Deductions, on the one hand, and observations, on the other, will be developed in two separate domains. If some passage is established from one to the other, it will be done in an artificial way. The legitimacy of such transitions will no longer be justifiable once the very principles of the theory have been deprived of all justification. Thus will we see the corollaries of a deduction being applied to the objects that the very axioms of this deduction declared nonexistent.

IX

The study of various electrical effects has led to assuming—then, it seems, to establishing—that within gases there are very small, very fast-moving electrically charged particles, given the name *electrons*. An electron acts in the fashion of an electrical current moving in a conducting body by quickly displacing in space the electrical charge it bears. The study of currents is a new chapter in electrodynamics. This chapter is still to be written.

To construct an electrodynamics for electrons, it was possible and desirable, it seems, to follow the prudent method Ampère, W. Weber, and Franz Neumann had used to construct the electrodynamics of conducting bodies. But this method required delicate experiments, penetrating intuitions, and arduous discussions, of which the works of W. Weber, Bernhard Riemann, and Clausius gave a first glimpse. It called for much ingenuity and much time. Algebraism found a means of proceeding with less bother and more haste. The intensity of previously known currents figured in Maxwell's equations. To these were added purely and simply the intensity of the *convection current* due to electron motion, without any change in the form of the equations; thus, one had the fundamental postulate of the new electrodynamics. As soon as a Dutch physicist, Lorentz, had proposed this hypothesis, German savants proceeded with great enthusiasm to deduce electron physics from it.

Thus, this physics rested completely on a simple generalization of Maxwell's equations, building on a beam known to be worm-eaten and therefore rendering the whole monument decrepit. Maxwell's equations contained a formal contradiction with the very existence of magnets and had not been cured of this vice when the convection current was introduced. The new electrodynamics presented itself at the outset as a set of corollaries of an inadmissible postulate.

This theory, vitiated by the very hypotheses that support it, did not hesitate, however, to pose as critic and reformer of doctrines considered the most solid until then. Rational mechanics, that elder sister of physical theories, which all the younger doctrines had until then taken as a guide and from which they frequently even tried to derive their principles, was itself shaken to its very foundations by this new arrival. In the name of electron physics, some proposed to renounce the principles of inertia and entirely to transform the notion of mass. That was needed so that the new doctrine would not be contradicted by the facts. Not for an instant did anyone ask if this contradiction, rather than requiring the overthrow of mechanics, did not signal the incorrectness of the hypotheses on which electron theory rested and did not mark the necessity of replacing or modifying them. These hypotheses were posited as postulates by the geometrical mind. It unfolded their consequences with an imperturbable assurance, triumphant in the very ruins which the conquering theory piled up among the old established doctrines. Guided, nevertheless, by the experience of the past, instructed by the history of great scientific progress, the intuitive mind suspected a poor indication of truth in this devastating march.

Moreover, by that inconsistency to which a reason deprived of intuition is so often condemned, the supporters of electron physics did not fail to use, in practice, the very theories condemned by the doctrine, when they were not unfolding the consequences of their preferred doctrine. Their deductions required the rejection of rational mechanics; but, without any scruples, they had recourse to the theorems of rational mechanics to interpret the readings of instruments whose information they borrowed.

X

The new physics was not content to enter into conflict with other physical theories in general, and with rational mechanics in particular. It did not recoil from contradicting common sense.

A delicate experiment in optics performed by M. Michelson happens to disagree with electron physics, as it does with most theories of optics proposed to this day. In that experiment, at least if it is duly confirmed and properly interpreted, the intuitive mind advises us to see the proof that, until now, no optical theory has been irreproachable and that it is necessary to revise every theory to some extent. But the geometrical mind of the German physicist holds another opinion. It has found a way to bring into agreement the equations of electron physics and the results of Michelson's experiment; in order to achieve this, it needed to overthrow the notions that common sense provides us regarding space and time.

The two notions of space and time appear to all men to be independent of one another. The new physics unites them with an indissoluble bond. The postulate securing this connection, which is truly an algebraic definition of time, has been called *the principle of relativity*. This principle of relativity, moreover, is so plainly a creation of the geometrical mind that one cannot give it a proper definition in ordinary language without having recourse to algebraic formulas.

At least one can show, by citing one of the consequences of the principle of relativity, the extent to which the link it establishes between the notions of space and time collides against the most formal assertions of common sense.

Our reason does not establish a necessary relation between the distance traversed by a mobile body and the time elapsed. However long a path may be, we can imagine it to be described in as small a time as we wish; however great a speed, we can always conceive a greater speed. No doubt, this greater speed could be unrealizable in fact; it could be that no physical means actually existed that is capable of launching a body with a speed greater than a given limit. But this impossibility, which seems to be imposed on the power of the engineer, would not present any insuperable absurdity to the thought of the theorist.

It is no longer the same if one admits the principle of relativity as it has been conceived by the likes of Einstein, Max Abraham, Minkowski, and von Laue: A body cannot move faster than the speed at which light is propagated in the void. And this impossibility is not a simple physical impossibility, one entailed as an effect by the absence of all means capable of producing it; it is a logical impossibility. For someone who holds to the principle of relativity, to speak of a speed surpassing the speed of light is to utter words devoid of meaning; it is to contradict the definition of time itself.

The fact that the principle of relativity is disconcerting to all the intuitions of common sense is not something that excites the suspicion of German physicists against it. To accept it is to upset all the doctrines that speak of space, time, and motion: all the theories of mechanics and physics.

Such devastation holds nothing unpleasant for German thought. The geometrical mind of the Germans will rejoice in reconstructing a whole physics based on the principle of relativity over the terrain it has just swept clean. If this new physics, disdainful of common sense, goes against all that observation and experiment have allowed us to construct in the domain of celestial and terrestrial mechanics, the purely deductive method will only be more proud of the inflexible rigor with which it will have followed the ruinous consequence of its postulate to its end.

Describing the "order of geometry," Pascal said:

> It does not define everything, and it does not prove everything [and in that respect it is inferior]. But it posits only those things which are clear and constant to natural light, and that is why it is perfectly true, inasmuch as nature supports it when other explanation fails.
>
> This orderly procedure, the most perfect known to men, consists not in defining everything and demonstrating everything, nor in defining or demonstrating nothing, but rather in maintaining a proper balance, by not defining those things that are clear and known to all men, and by defining all others; by not proving those things that are known to all men and by proving others. This order is sinned against equally by those who undertake to define and prove everything and by those who neglect doing so in matters that are not self-evident.
>
> This is what geometry teaches perfectly. It does not define any such things as space, time, motion, number, equality, nor very numerous similar things. . . .
>
> It may perhaps be considered strange that geometry cannot define any of the things with which it is principally concerned, for it can define neither motion, nor number, nor space. And yet these three things are the ones with which it deals particularly. . . . But we should not be surprised if we observe that this admirable science, in dealing only with the simplest things, makes them incapable of definition because of this very quality of simplicity which makes them worthy of being the objects of that science. Thus the lack of definition is the mark of excellence rather than a defect, because it is not due to any obscurity in these things but rather to their extreme clarity which is such that, although it does not carry the conviction of demonstration, nevertheless it does afford the same full certainty.[4]

The exclusively geometrical mind does not want to concede to the intuitive mind the power of drawing from common sense certain knowledge

4. [Pascal, "The Mind of the Geometrician," in *Great Shorter Works*, trans. A. Cailliet and J. C. Blankenagel (Westport: Greenwood Press, 1974), pp. 191–195.]

contained in it, endowed with that extreme evidence which does not have the conviction of demonstration but has all its certainty. It knows no other evidence and no other certainty than that of definitions and demonstrations, so that it comes to dream of a science in which all propositions have been demonstrated. And since it is contradictory to define everything and demonstrate everything, it wishes at least to reduce all undefined notions and undemonstrated judgments to the smallest number possible. The only ideas it consents to receive without definition are the ideas of whole number, equality, inequality, and the addition of whole numbers; the only propositions it readily receives without requiring a demonstration are the axioms of arithmetic. When it has developed the full doctrine of algebra from such notions and principles as these, it understands well how to reduce all of science to nothing but a chapter of that algebra. The ideas of space, time, and motion are presented to us by common knowledge as simple and irreducible ideas, which cannot be reconstructed with the aid of operations bearing on whole numbers. They are, therefore, essentially incapable of algebraic definition. But that is no obstacle! The geometrical mind refuses to consider the space, time, and motion which all people conceive clearly and about which they can talk among themselves without ever ceasing to understand one another. By operations referring to algebraic expressions—that is, in the last analysis, to whole numbers—it fabricates for itself its own space, its own time, its own motion. It subjects this space, time, and motion to postulates which are arbitrarily arranged algebraic equations. And when it has rigorously deduced a long series of theorems from these definitions and postulates, according to the rules of calculation, it says it has produced a geometry, a mechanics, a physics, although it has only developed chapters of algebra. That is how Riemann's geometry was constructed; that is how relativity physics was constructed; that is how German science progresses, proud of its algebraic rigidity, looking with scorn on the good sense of which all people have received a share.

XI

Of that German science, we have still considered only geometry, mechanics, and physics. They are the parts that use mathematics constantly and, therefore, those that most easily take on the algebraic form. But we believe that somewhat attentive observers will also find the characteristics we have recognized when examining these various chapters of German science if they examine the other chapters.

No one is unaware, for example, of the extraordinary development that the study of chemistry has undergone in Germany. Now, the rise of German chemistry dates from the day that atomic notation issued from the notions of chemical type and valence, notions given birth by the works of the likes of J.-B. Dumas, Laurent, Gerhardt, Williamson, and Wurtz. This notation, in fact, allows one to predict, enumerate, and classify the reactions, syntheses, and isomerisms of carbon compounds with the help of rules furnished by the part of algebra called *analysis situs* [topology]. Thus, it is the study of carbon compounds—that is, organic chemistry— henceforth subject to the control of the geometrical mind—that has produced innumerable surges of extraordinary vigor in German laboratories. In contrast, mathematical operations of atomic notation are of restricted usage in the numerous chapters that make up inorganic chemistry. The intuitive mind is still the instrument that unscrambles the complexity of reactions and classifies compounds. Thus, these chapters of chemistry have not received from German science a tribute comparable to the one paid to them by French science.

We do not wish to venture into the domain of criticism and history; *ne sutor ultra crepidam* [Let the cobbler stick to his last]. It seems, however, to our outsider's eyes, that one could find occasion to make similar remarks for these.

According to the taste of French science, historical studies essentially arise from the intuitive mind. The ingenuity and vivid imagination belonging to the French perhaps carried them too often to adventuresome conclusions and syntheses of fantasy. By extolling minute research into sources and the patient verification of texts, and by requiring the production of solid documents to support the least assertion, the geometrical mind of the Germans has come, quite happily, to restrain the rashness of an overly impulsive intuitive mind. But it has not contented itself with reminding the latter that its power would become too fragile if it did not support its intuitions with the help of certain proofs; it has wanted to exclude it entirely from studies in which, until now, it had reigned supreme. We have therefore seen develop that German erudition whose method, regulated like clockwork, claimed to lead us from texts to conclusions by infallible paths "without the least appeal to judgment and good, common sense." By the rigor of its proceedings, by the systematic look of its operations, even by the form of its language—unintelligible to the outsider—and the signs it is often happy to use, this erudition visibly strove to copy the look of mathematical analysis.

Now, studies that require critical sense are precisely those in which the absolute and rigid method of algebra is to the greatest extent out of place.

It is above all with respect to the examination of a historical text that one can say with Pascal:

> Principles are in ordinary usage and there for all to see. There is no need to turn our heads, or strain ourselves; it is only a question of good sight, but it must be good; for the principles are so intricate and numerous that it is almost impossible not to miss some. Now the omission of one principle can lead to error, and so one needs very clear sight to see all the principles as well as an accurate mind to avoid drawing false conclusions from known principles.[5]

In order to retain a clear view of these numerous principles that "are in ordinary usage and there for all to see," is it reasonable to put the inextricable and tightly woven mesh of German method between the eye of good sense and the documents one asks it to read?

XII

Should we give a conclusion to these various reflections? A conclusion seems to follow so naturally from what has gone before that we feel somewhat diffident about formulating it; thus, we shall do it very briefly.

French science and German science both deviate from ideal and perfect science, but they deviate in two opposite ways. The one possesses excessively what the other possesses meagerly. In one, the geometrical mind reduces the intuitive mind to the point of suffocation. In the other, the intuitive mind dispenses too readily with the geometrical mind.

Therefore, in order for human science to develop in its fullness and subsist in harmonious equilibrium, it is good to see French science and German science flourishing side by side without trying to supplant each other. Each of them ought to understand that it finds in the other its indispensable complement.

Therefore, the French will always find profit in pondering the works of German scholars. They will encounter there either the solid proof of truths they discovered and formulated before being fully certain of them, or the refutation of the errors which a hasty intuition caused them to accept.

It will always be of use to Germans to study the writings of French inventors. They will find there, so to speak, the statement of problems that their patient analysis should apply itself toward resolving. They will hear

5. [B. Pascal, *Pensées*, 512 (1), trans. Krailshimer, p. 211.]

there the protestations of good sense against the excess of their geometrical mind.

I think no one from the other side of the Rhine would dare to deny that German science in the nineteenth century took its departure from the work of great French thinkers; and no one from this side would dream of failing to recognize the contributions with which, later on, German science has enriched our mathematics, physics, chemistry, and history.

These two sciences, then, ought to retain harmonious relations with each other. It does not follow that they should be placed at the same rank. Intuition discovers truths; demonstration comes after and certifies them. The geometrical mind gives body to the edifice the intuitive mind conceived first. There is a hierarchy between these two minds that is similar to the one that orders the mason with respect to the architect. Masons do useful work only if their work conforms to the architect's plan. The geometrical mind does not pursue fruitful deductions if it does not direct them toward the end the intuitive mind has discerned.

On the other hand, the geometrical mind can truly certify a rigor without reproach for the part of science constructed by the deductive method. But the rigor of science is not its truth. The intuitive mind alone judges whether the principles of deduction are acceptable, whether the consequences of demonstration are in conformity with reality. For science to be true, it is not sufficient that it be rigorous; it must start from good sense in order to end up with good sense.

The geometrical mind that inspires German science confers on it the force of a perfect discipline. But this narrowly disciplined method can only lead to disastrous results if it continues to put itself under the orders of an arbitrary and foolish algebraic imperialism. If it wishes to do useful and beautiful work, it must receive the order it obeys from the principal depository of good sense in the world—that is, from French science. *Scientia germanica ancilla scientiae gallicae* [German science is the handmaiden of French science].

Index

Compiled with the assistance of Maureen A. McCormick

Duhem
Essays in the History and Philosophy of Science

Here, for the first time in English, are the philosophical essays—including the first statement of the "Duhem Thesis"—that formed the basis for *Aim and Structure of Physical Theory*, together with new translations of the historiographical essays presenting the equally celebrated "Continuity Thesis" by Pierre Duhem (1861–1916), a founding figure of the history and philosophy of science. Prefaced by an introduction on Duhem's intellectual development and continuing significance, here as well are important subsequent essays in which Duhem elaborated key concepts and critiqued such contemporaries as Henri Poincaré and Ernst Mach. Together, these works offer a lively picture of the state of science at the turn of the century while addressing methodological issues that remain at the center of debate today.

ROGER ARIEW is Professor of Philosophy, Virginia Polytechnic Institute and State University.
PETER BARKER is Professor of Philosophy, University ⊂ Oklahoma, Norman.

9 780872 203082

ISBN 0-87220-308-5